Applied Machine Learning and High-Performance Computing on AWS

Accelerate the development of machine learning applications following architectural best practices

Mani Khanuja

Farooq Sabir

Shreyas Subramanian

Trenton Potgieter

BIRMINGHAM—MUMBAI

Applied Machine Learning and High-Performance Computing on AWS

Publishing Product Manager: Dhruv Jagdish Kataria
Senior Editor: Nathanya Dias
Content Development Editor: Manikandan Kurup
Technical Editor: Devanshi Ayare
Copy Editor: Safis Editing
Project Coordinator: Farheen Fathima
Proofreader: Safis Editing
Indexer: Rekha Nair
Production Designer: Shankar Kalbhor
Marketing Coordinator: Shifa Ansari

First published: December 2022

Production reference: 3100423

Published by Packt Publishing Ltd.
Livery Place
35 Livery Street
Birmingham
B3 2PB, UK.

ISBN 978-1-80323-701-5

www.packtpub.com

Contributors

About the authors

Mani Khanuja is a seasoned IT professional with over 17 years of software engineering experience. She has successfully led machine learning and artificial intelligence projects in various domains, such as forecasting, computer vision, and natural language processing. At AWS, she helps customers to build, train, and deploy large machine learning models at scale. She also specializes in data preparation, distributed model training, performance optimization, machine learning at the edge, and automating the complete machine learning life cycle to build repeatable and scalable applications.

Farooq Sabir is a research and development expert in machine learning, data science, big data, predictive analytics, computer vision, and image and video processing. He has over 10 years of professional experience.

Shreyas Subramanian helps AWS customers build and fine-tune large-scale machine learning and deep learning models, and rearchitect solutions to help improve the security, scalability, and efficiency of machine learning platforms. He also specializes in setting up massively parallel distributed training, hyperparameter optimization, and reinforcement learning solutions, and provides reusable architecture templates to solve AI and optimization use cases.

Trenton Potgieter is an expert technologist with 25 years of both local and international experience across multiple aspects of an organization; from IT to sales, engineering, and consulting, on the cloud and on-premises. He has a proven ability to analyze, assess, recommend, and design appropriate solutions that meet key business criteria, as well as present and teach them from engineering to executive levels.

About the reviewers

Ram Vittal has over 20 years of experience engineering software solutions for solving complex challenges across various business domains. Ram started his career with mainframes and then moved on to building distributed systems using Java technologies. Ram started his cloud journey in 2015 and has helped enterprise customers migrate, optimize, and scale their workloads on AWS. As a principal machine learning solutions architect, Ram has helped customers solve challenges across areas such as security, governance, and big data for building machine learning platforms. Ram has delivered thought leadership on big data, machine learning, and cloud strategies. Ram holds 11 AWS certifications and has a master's degree in computer engineering.

Chakravarthy Nagarajan is a technical evangelist with 21 years of industry experience in machine learning, big data, and high-performance computing. He is currently working as a Principal AI/ML Specialist Solutions Architect at AWS in Bay Area, USA. He helps customers to solve real-world complex business problems by building prototypes with end-to-end AI/ML solutions on cloud, and edge devices. His specialization includes computer vision, natural language processing, time series forecasting, and personalization. He is also a public speaker and has published multiple blogs and white papers on HPC and AI/ML. On the academic front, he completed his MBA from United Institute, Brussels, and has multiple certifications in AI and ML.

Anna Astori holds a master's degree in computational linguistics and artificial intelligence from Brandeis University. Over the years, Anna has worked on multiple large-scale machine learning and data science applications for companies such as Amazon and Decathlon. Anna is an AWS Certified Developer and Solutions Architect. She speaks at conferences and podcasts, reviews talk proposals for tech conferences, and writes about Python and machine learning for curated publications on Medium. She is currently a co-director of the Women Who Code Boston network.

Kevin Sayers is a research scientist based in Colorado with expertise in **high-performance computing (HPC)**. He holds a master's degree in bioinformatics. His work has primarily focused on scientific workflows and he is an open source contributor to a number of workflow tools. He has previously worked with university HPC centers and a national lab supporting the HPC user community. He currently works in cloud HPC bringing customer HPC and machine learning workloads to the cloud.

Table of Contents

Part 1: Introducing High-Performance Computing

1

2

Part 2: Applied Modeling

5

Data Analysis   95

6

Distributed Training of Machine Learning Models   127

7

Deploying Machine Learning Models at Scale   151

8

Optimizing and Managing Machine Learning Models for Edge Deployment   179

Part 3: Driving Innovation Across Industries

Preface

Applied **machine learning** (**ML**) and **high-performance computing** (**HPC**) have been integral to tackling the world's most complex problems, including genomics, autonomous vehicles, computational fluid dynamics, and numerical optimization. Due to the high computing power required for these applications, HPC was once only used by large organizations that could afford it. Now, it is available for use by everyone, from individual research groups to start-ups and big enterprises, using **Amazon Web Services** (**AWS**) cloud technology.

This book provides a complete step-by-step explanation of the essential concepts with practical examples. You will begin by exploring virtually unlimited infrastructure and fast networking for scalable HPC on AWS, including an overview of the relevant tools and technologies. You'll learn how to develop large-scale ML applications using HPC on AWS, you will understand the various architectural components, and you'll learn about performance optimization, with hands-on application to real-world use cases in various domains.

By the end of this book, you will be able to build and deploy your own large-scale ML applications using HPC on AWS, following the industry best practices and addressing the key pain points encountered in the application life cycle.

Who this book is for

This book is designed for scientists, researchers, engineers, architects, and managers in fields such as ML, **artificial intelligence** (**AI**), numerical optimization, and genomics, which require HPC. This book is ideal for people working in the following roles:

- Data and AI scientists
- AI/ML engineers
- AI/ML product managers
- AI product owners
- AI/ML researchers

In general, any ML enthusiast with a foundational knowledge of Python will be able to read, understand, and apply the knowledge gained from this book.

What this book covers

Chapter 1, High-Performance Computing Fundamentals, introduces the concepts of HPC, highlighting the importance of HPC as it relates to real-world scenarios. We then talk about technological advancements in HPC and how you can use it to solve complex business problems with unlimited capacity, the most advanced computing capabilities, and the elasticity of the cloud, while still optimizing cost, to innovate faster and gain a competitive business advantage.

Chapter 2, Data Management and Transfer, dives into data management and transfer. The first step to running HPC applications on the cloud is to move the required data to the cloud. Therefore, this chapter focuses on different aspects of data migration, including challenges and key pain points businesses might face, and how AWS data migration services can help resolve them, following the best practices and still maintaining data integrity, consistency and security.

Chapter 3, Compute and Networking, explains how once you have the data in the cloud, you need to understand the compute options provided by AWS as well as their differences, in order to optimally select the right option based on your business requirements. Furthermore, to scale and secure your HPC applications, we then dive into the Networking section, to explain the concepts of private VPC, low latency networking, and optimizing the performance of inter-instance communications.

Chapter 4, Data Storage, explains that before you can begin performing ML using HPC, it's important to understand the storage options and storage costs for both the transient and permanent storage requirements. This chapter dives deep into the various storage services in the AWS ecosystem, to help you select the right tool for the right job.

Chapter 5, Data Analysis, teaches you how to explore the data, collect metrics, perform data correlation, and process large amounts of data using AWS to ensure data quality before using it for training ML models.

Chapter 6, Distributed Training of Machine Learning Models, shows you how to implement a large ML model using vast amounts of data by leveraging distributed data parallel and model parallel concepts.

Chapter 7, Deploying Machine Learning Models at Scale, discusses model deployment and inference. We will start with what managed deployment means on AWS, then go on to discuss the right deployment options, followed by various inference options (batch, asynchronous, and real-time). We will then discuss the reliability and availability of model endpoints on AWS infrastructure and the blue/green deployment option for different versions of the model.

Chapter 8, Optimizing and Managing Machine Learning Models for Edge Deployment, explores ML models on edge devices. We will start with an introduction to edge computing, followed by factors we need to consider for the optimization of machine learning models. You will also learn about architecture design for edge deployment.

Chapter 9, Performance Optimization for Real-Time Inference, discusses some of the key performance metrics used for ML models, techniques to reduce the memory footprint of large models, choosing the right machine (instance type) to deploy the model, load testing, and performance tuning of models.

Chapter 10, Data Visualization, covers Amazon SageMaker Data Wrangler, a tool that enables users working in the domains of data science, ML, and analytics to build insightful data visualizations without writing much code. In addition, we will also briefly touch upon the topic of AWS's graphics-optimized instances, since these instances can be used to create animated live data visualizations along with other high-performance computing applications such as game streaming, and ML.

Chapter 11, Computational Fluid Dynamics, introduces the field of **computational fluid dynamics** (**CFD**), which uses numerical analysis to solve fluid flow problems. CFD has far-reaching applications in many industries such as the automotive industry, oil and gas, and manufacturing. We will discuss how CFD solvers can be used on AWS for massive-scale problems, and how recent advances in ML help with accelerating CFD applications.

Chapter 12, Genomics, introduces the field of genomics, and how AWS can help customers with large-scale genomics applications that typically use large datasets. We also discuss typical architectures for storing and analyzing such data, along with how ML is currently being applied to genomics, followed by an example of protein structure analysis.

Chapter 13, Autonomous Vehicles, discusses **autonomous vehicles** (**AVs**) and the technology used to safely and efficiently operate a vehicle at various levels of automation. Today, companies use petabytes of data from sensors and cameras on large-scale clusters to perform **deep neural network** (**DNN**) training. Specifically, we discuss services that support AV development, architectures for large-scale data processing, and the use of DNNs for training AV models.

Chapter 14, Numerical Optimization, introduces you to what numerical optimization is, and why it is important to solve large-scale problems that we might encounter in this space. We will touch upon some common use cases in this domain and the HPC available to solve these use cases. We will also discuss the application of ML to numerical optimization.

To get the most out of this book

To follow the text and examples in this book, you are recommended to have a foundational knowledge of Python and HPC, and an intermediate understanding of data analysis, ML, and AI. In addition, you should have access to the following technology tools to work through the code and experimentation examples:

Software/hardware requirements	Operating system requirements
Web browser (Chrome, Firefox, or Safari)	Windows, Linux, or macOS
AWS account	

If you are using the digital version of this book, we advise you to type the code yourself or access the code from the book's GitHub repository (a link is available in the next section). Doing so will help you avoid any potential errors related to the copying and pasting of code.

Download the example code files

You can download the example code files for this book from GitHub at https://github.com/PacktPublishing/Applied-Machine-Learning-and-High-Performance-Computing-on-AWS. If there's an update to the code, it will be updated in the GitHub repository.

We also have other code bundles from our rich catalog of books and videos available at https://github.com/PacktPublishing/. Check them out.

Download the color images

We also provide a PDF file that has color images of the screenshots and diagrams used in this book. You can download it here: https://packt.link/SqXiF.

Conventions used

There are a number of text conventions used throughout this book.

Code in text: Indicates code words in text, database table names, folder names, filenames, file extensions, pathnames, dummy URLs, user input, and Twitter handles. Here is an example: "The next step is to use the model object's deploy() method to create an HTTPS endpoint."

A block of code is set as follows:

```
import numpy as np
import json

with open("horse_cart.jpg", "rb") as f:
    payload = f.read()
    payload = bytearray(payload)
```

When we wish to draw your attention to a particular part of a code block, the relevant lines or items are set in bold:

```
smp_options = {
    "enabled":True,
    "parameters": {
        "partitions": 1,
        "placement_strategy": "spread",
        "pipeline": "interleaved",
```

```
        "optimize": "speed",
        "ddp": True,
    }
}
```

Any command-line input or output is written as follows:

```
$ sudo tail -f /greengrass/v2/logs/com.greengrass.
SageMakerEdgeManager.ImageClassification.log
```

Bold: Indicates a new term, an important word, or words that you see onscreen. For instance, words in menus or dialog boxes appear in **bold**. Here is an example: "On the **Configure components** page, select the **aws.greengrass.SageMakerEdgeManager** component, and choose **Configure component**."

> **Tips or important notes**
> Appear like this.

Get in touch

Feedback from our readers is always welcome.

General feedback: If you have questions about any aspect of this book, email us at customercare@ packtpub.com and mention the book title in the subject of your message.

Errata: Although we have taken every care to ensure the accuracy of our content, mistakes do happen. If you have found a mistake in this book, we would be grateful if you would report this to us. Please visit www.packtpub.com/support/errata and fill in the form.

Piracy: If you come across any illegal copies of our works in any form on the internet, we would be grateful if you would provide us with the location address or website name. Please contact us at copyright@packt.com with a link to the material.

If you are interested in becoming an author: If there is a topic that you have expertise in and you are interested in either writing or contributing to a book, please visit authors.packtpub.com.

Share Your Thoughts

Once you've read *Applied Machine Learning and High-Performance Computing on AWS,* we'd love to hear your thoughts! Scan the QR code below to go straight to the Amazon review page for this book and share your feedback.

https://packt.link/r/1-803-23701-5

Your review is important to us and the tech community and will help us make sure we're delivering excellent quality content.

Download a free PDF copy of this book

Thanks for purchasing this book!

Do you like to read on the go but are unable to carry your print books everywhere? Is your eBook purchase not compatible with the device of your choice?

Don't worry, now with every Packt book you get a DRM-free PDF version of that book at no cost.

Read anywhere, any place, on any device. Search, copy, and paste code from your favorite technical books directly into your application.

The perks don't stop there, you can get exclusive access to discounts, newsletters, and great free content in your inbox daily

Follow these simple steps to get the benefits:

1. Scan the QR code or visit the link below

https://packt.link/free-ebook/978-1-80323-701-5

2. Submit your proof of purchase
3. That's it! We'll send your free PDF and other benefits to your email directly

Part 1: Introducing High-Performance Computing

The objective of *Part 1* is to introduce the concepts of **high-performance computing** (**HPC**) and the art of possibility, and, most importantly, to make you understand that what was once confined to large enterprises, government bodies, or academic institutions is now reachable, as well as how industries are leveraging it to drive innovation.

This part comprises the following chapters:

- *Chapter 1, High-Performance Computing Fundamentals*
- *Chapter 2, Data Management and Transfer*
- *Chapter 3, Compute and Networking*
- *Chapter 4, Data Storage*

High-Performance Computing Fundamentals

High-Performance Computing (HPC) impacts every aspect of your life, from your morning coffee to driving a car to get to the office, knowing the weather forecast, your vaccinations, movies that you watch, flights that you take, games that you play, and many other aspects. Many of our actions leave a digital footprint, leading to the generation of massive amounts of data. In order to process such data, we need a large amount of processing power. HPC, also known as *accelerated computing*, aggregates the computing power from a cluster of nodes and divides the work among various interconnected processors, to achieve much higher performance than could be achieved by using a single computer or machine, as shown in *Figure 1.1*. This helps in solving complex scientific and engineering problems, in critical business applications such as drug discovery, flight simulations, supply chain optimization, financial risk analysis, and so on:

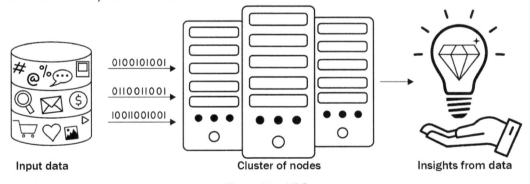

Input data   Cluster of nodes   Insights from data

Figure 1.1 – HPC

For example, drug discovery is a data-intensive process, which involves computationally heavy calculations to simulate how virus protein binds with human protein. This is an extremely expensive process and may take weeks or months to finish. With the unification of **Machine Learning** (ML) with accelerated computing, researchers can simulate drug interactions with protein with higher speed and accuracy. This leads to faster experimentation and significantly reduces the time to market.

In this chapter, we will learn the fundamentals and importance of HPC, followed by technological advancements in the area. We will understand the constraints and how developers can benefit from the elasticity of the cloud while still optimizing costs to innovate faster to gain a competitive business advantage.

In this chapter, we will cover the following topics:

- Why do we need HPC?
- Limitations of on-premises HPC
- Benefits of doing HPC on the cloud
- Driving innovation across industries with HPC

Why do we need HPC?

According to Statista, the rate of growth of data globally is forecast to increase rapidly and reached 64.2 zettabytes in 2020. By 2025, the volume of data is estimated to grow to more than 180 zettabytes. Due to the COVID-19 pandemic, data growth in 2020 reached a new high as more people were learning online and working remotely from home. As data is continuously increasing, the need to be able to analyze and process it also increases. This is where HPC is a useful mechanism. It helps organizations to think beyond their existing capabilities and explore possibilities with advanced computing technologies. Today HPC applications, which were once confined to large enterprises and academia, are trending across a wide range of industries. Some of these industries include material sciences, manufacturing, product quality improvement, genomics, numerical optimization, computational fluid dynamics, and many more. The list of applications for HPC will continue to increase, as cloud infrastructure is making it accessible to more organizations irrespective of their size, while still optimizing cost, helping to innovate faster and gain a competitive advantage.

Before we take a deeper look into doing HPC on the cloud, let's understand the limitations of running HPC applications on-premises, and how we can overcome them by using specialized HPC services provided by the cloud.

Limitations of on-premises HPC

HPC applications are often based on complex models trained on a large amount of data, which require high-performing hardware such as **Graphical Processing Units (GPUs)** and software for distributing the workload among different machines. Some applications may need parallel processing while others may require low-latency and high-throughput networking. Similarly, applications such as gaming and video analysis may need performance acceleration using a fast input or output subsystem and GPUs. Catering to all of the different types of HPC applications on-premises might be daunting in terms of cost and maintenance.

Some of the well-known challenges include, but are not limited to, the following:

- High upfront capital investment
- Long procurement cycles
- Maintaining the infrastructure over its life cycle
- Technology refreshes
- Forecasting the annual budget and capacity requirement

Due to the above-mentioned constraints, planning for an HPC system can be a grueling process, **Return On Investment** (**ROI**) for which might be difficult to justify. This can be a barrier to innovation, with slow growth, reduced efficiency, lost opportunities, and limited scalability and elasticity. Let's understand the impact of each of these in detail.

Barrier to innovation

The constraints of on-premises infrastructure can limit the system design, which will be more focused on the availability of the hardware instead of the business use case. You might not consider some new ideas if they are not supported by the existing infrastructure, thus obstructing your creativity and hindering innovation within the organization.

Reduced efficiency

Once you finish developing the various components of the system, you might have to wait in long prioritized queues to test your jobs, which might take weeks, even if it takes only a few hours to run. On-premises infrastructure is designed to capitalize on the utilization of expensive hardware, often resulting in very convoluted policies for prioritizing the execution of jobs, thus decreasing your productivity and ability to innovate.

Lost opportunities

In order to take full advantage of the latest technology, organizations have to refresh their hardware. Earlier, the typical refresh cycle of three years was enough to stay current, to meet the demands of HPC workloads. However, due to fast technological advancements and a faster pace of innovation, organizations need to refresh their infrastructure more often, otherwise, it might have a larger downstream business impact in terms of revenue. For example, technologies such as **Artificial Intelligence** (**AI**), ML, data visualization, risk analysis of financial markets, and so on, are pushing the limits of on-premises infrastructure. Moreover, due to the advent of the cloud, a lot of these technologies are cloud native, and deliver higher performance on large datasets when running in the cloud, especially with workloads that use transient data.

Limited scalability and elasticity

HPC applications rely heavily on infrastructure elements such as containers, GPUs, and serverless technologies, which are not readily available in an on-premises environment, and often have a long procurement and budget approval process. Moreover, maintaining these environments, making sure they are fully utilized, and even upgrading the OS or software packages, requires skills and dedicated resources. Deploying different types of HPC applications on the same hardware is very limiting in terms of scalability and flexibility and does not provide you with the right tools for the job.

Now that we understand the limitations of doing HPC on-premises, let's see how we can overcome them by running HPC workloads on the cloud.

Benefits of doing HPC on the cloud

With virtually unlimited capacity on the cloud, you can move beyond the constraints of on-premises HPC. You can reimagine new approaches based on the business use case, experiment faster, and gain insights from large amounts of data, without the need for costly on-premises upgrades and long procurement cycles. You can run complex simulations and deep learning models in the cloud and quickly move from idea to market using scalable compute capacity, high-performance storage, and high-throughput networking. In summary, it enables you to drive innovation, collaborate among distributed teams, improve operational efficiency, and optimize performance and cost. Let's take a deeper look into each of these benefits.

Drives innovation

Moving HPC workloads to the cloud, helps you break barriers to innovation, and opens the door for unlimited possibilities. You can quickly fail forward, try out thousands of experiments, and make business decisions based on data. The benefit that I really like is that, once you solve the problem, it remains solved and you don't have to revisit it after a system upgrade or a technology refresh. It eliminates reworking and the maintenance of hardware, lets you focus on the business use case, and enables you to quickly design, develop, and test new products. The elasticity offered by the cloud, allows you to grow and shrink the infrastructure as per the requirements. Additionally, cloud-based services offer native features, which remove the heavy lifting and let you adopt tested and verified HPC applications, without having to write and manage all the utility libraries on your own.

Enables secure collaboration among distributed teams

HPC workloads on the cloud allow you to share designs, data, visualizations, and other artifacts globally with your teams, without the need to duplicate or proliferate sensitive data. For example, building a digital twin (a real-time digital counterpart of a physical object) can help in predictive maintenance. It can get the state of the object in real time and it monitors and diagnoses the object (asset) to optimize its performance and utilization. To build a digital twin, a cross-team skill set is needed, which might be remotely located to capture data from various IoT sensors, performing extensive what-if analysis and meticulously building a simulation model to develop an accurate representation of the physical object. The cloud provides a collaboration platform, where different teams can interact with a simulation model in near real time, without moving or copying data to different locations, and ensures compliance with rapidly changing industry regulations. Moreover, you can use native features and services offered by the cloud, for example, AWS IoT TwinMaker, which can use the existing data from multiple sources, create virtual replicas of physical systems, and combine 3D models to give you a holistic view of your operations faster and with less effort. With a broad global presence of HPC technologies on the cloud, it allows you to work together with your remote teams across different geographies without trading off security and cost.

Amplifies operational efficiency

Operational efficiency means that you are able to support the development and execution of workloads, gain insights, and continuously improve the processes that are supporting your applications. The design principles and best practices include automating processes, making frequent and reversible changes, refining your operations frequently, and being able to anticipate and recover from failures. Having your HPC applications on the cloud enables you to do that, as you can version control your infrastructure as code, similar to your application code, and integrate it with your **Continuous Integration** and **Continuous Delivery (CI/CD)** pipelines. Additionally, with on-demand access to unlimited compute capacity, you will no longer have to wait in long queues for your jobs to run. You can skip the wait and focus on solving business critical problems, providing you with the right tools for the right job.

Optimizes performance

Performance optimization involves the ability to use resources efficiently and to be able to maintain them as the application changes or evolves. Some of the best practices include making the implementation easier for your team, using serverless architectures where possible, and being able to experiment faster. For example, developing ML models and integrating them into your application requires special expertise, which can be alleviated by using out-of-the-box models provided by cloud vendors, such as services in the AI and ML stack by AWS. Moreover, you can leverage the compute, storage, and networking services specially designed for HPC and eliminate long procurement cycles for specialized hardware. You can quickly carry out benchmarking or load testing and use that data to optimize your workloads without worrying about cost, as you only pay for the amount of time you are using the resources on the cloud. We will understand this concept more in *Chapters 5, Data Analysis*, and *Chapter 6, Distributed Training of Machine Learning Models*.

Optimizes cost

Cost optimization is a continuous process of monitoring and improving resource utilization over an application's life cycle. By adopting the pay-as-you-go consumption model and increasing or decreasing the usage depending on the business needs, you can achieve potential cost savings. You can quickly commission and decommission HPC clusters in minutes, instead of days or weeks. This lets you gain access to resources rapidly, as and when needed. You can measure the overall efficiency by calculating the business value achieved and the cost of delivery. With this data, you can make informed decisions as well as understanding the gains from increasing the application's functionality and reducing cost.

Running HPC in the cloud helps you overcome the limitations associated with traditional on-premises infrastructure: fixed capacity, long procurement cycles, technology obsolescence, high upfront capital investment, maintaining the hardware, and applying regular **Operating System** (**OS**) and software updates. The cloud gives you unlimited HPC capacity virtually, with the latest technology to promote innovation, which helps you design your architecture based on business needs instead of available hardware, minimizes the need for job queues, and improves operational and performance efficiency while still optimizing cost.

Next, let's see how different industries such as **Autonomous Vehicles** (**AVs**), manufacturing, media and entertainment, life sciences, and financial services are driving innovation with HPC workloads.

Driving innovation across industries with HPC

Every industry and type of HPC application poses different kinds of challenges. The HPC solutions provided by cloud vendors such as AWS help all companies, irrespective of their size, which leads to emerging HPC applications such as reinforcement learning, digital twins, supply chain optimization, and AVs.

Let's take a look at some of the use cases in life sciences and healthcare, AV, and supply chain optimization.

Life sciences and healthcare

In life sciences and the healthcare domain, a large amount of sensitive and meaningful data is captured almost every minute of the day. Using HPC technology, we can harness this data to gain meaningful insights into critical diseases to save lives by reducing the time taken testing lab samples, drug discovery, and much more, as well as meeting the core security and compliance requirements.

The following are some of the emerging applications in the healthcare and life sciences domain.

Genomics

You can use cloud services provided by AWS to store and share genomic data securely, which helps you build and run predictive or real-time applications to accelerate the journey from genomic data to genomic insights. This helps to reduce data processing times significantly and perform casual analysis of critical diseases such as cancer and Alzheimer's.

Imaging

Using computer vision and data integration services, you can elevate image analysis and facilitate long-term data retention. For example, by using ML to analyze MRI or X-ray scans, radiology companies can improve operational efficiency and quickly generate lab reports for their patients. Some of the technologies provided by AWS for imaging include Amazon EC2 GPU instances, AWS Batch, AWS ParallelCluster, AWS DataSync, and Amazon SageMaker, which we will discuss in detail in subsequent chapters.

Computational chemistry and structure-based drug design

Combining the state-of-the-art deep learning models for protein classification, advancements in protein structure solutions, and algorithms for describing 3D molecular models with HPC computing resources, allows you to grow and reduce the time to market drastically. For example, in a project performed by Novartis on the AWS cloud, where they were able to screen 10 million compounds against a common cancer target in less than a week, based on their internal calculations, if they had performed a similar experiment in-house, then it would have resulted in about a $40 million investment. By running this experiment on the cloud using AWS services and features, they were able to use their 39 years of computational chemistry data and knowledge. Moreover, it only took 9 hours and $4,232 to conduct the experiment, hence increasing their pace of innovation and experimentation. They were able to successfully identify three out of the ten million compounds, that were screened.

Now that we understand some of the applications in the life sciences and healthcare domain, let us discuss how the automobile and transport industry is using HPC for building AVs.

AVs

The advancement of deep learning models such as reinforcement learning, object detection, and image segmentation, as well technological advancements in compute technology, and deploying models on edge devices, have paved the way for AV. In order to design and build an AV, all the components of the system have to work in tandem, including planning, perception, and control systems. It also requires collecting and processing massive amounts of data and using it to create a feedback loop, for vehicles to adjust their state based on the changing condition of the traffic on roads in real time. This entails having high I/O performance, networking, specialized hardware coprocessors such as GPUs or **Field Programmable Gate Arrays** (**FPGAs**), as well as analytics and deep learning frameworks. Moreover, before an AV can even start testing on actual roads, it has to undergo millions of miles of simulation to demonstrate safety performance, due to the high dimensionality of the environment, which is complex and time-consuming. By using the AWS cloud's virtually unlimited compute and storage capacity, support for advanced deep learning frameworks, and purpose-built services, you can drive a faster time to market. For example, In 2017, Lyft, an American transportation company, launched its AV division. To enhance the performance and safety of its system, it uses petabytes of data collected from its AV fleet to execute millions of simulations every year, which involves a lot of compute power. To run these simulations at a lower cost, they decided to take advantage of unused compute capacity on the AWS cloud, by using Amazon EC2 Spot Instances, which also helped them to increase their capacity to run the simulations at this magnitude.

Next, let us understand supply chain optimization and its processes!

Supply chain optimization

Supply chains are worldwide networks of manufacturers, distributors, suppliers, logistics, and e-commerce retailers that work together to get products from the factory to the customer's door without delays or damage. By enabling information flow through these networks, you can automate decisions without any human intervention. The key attributes to consider are real-time inventory forecasts, end-to-end visibility, and the ability to track and trace the entire production process with unparalleled efficiency. Your teams will no longer have to handle the minuscular details associated with supply chain decisions. With automation and ML you can resolve bottlenecks in product movement. For example, in the event of a pandemic or natural disaster, you can quickly divert goods to alternative shipping routes without affecting their on-time delivery.

Here are some examples of using ML to improve supply chain processes:

- **Demand Forecasting**: You can combine a time series with additional correlated data, such as holidays, weather, and demographic events, and use deep learning models such as DeepAR to get more accurate results. This will help you to meet variable demand and avoid over-provisioning.

- **Inventory Management**: You can automate inventory management using ML models to determine stock levels and reduce costs by preventing excess inventory. Moreover, you can use ML models for anomaly detection in your supply chain processes, which can help you in optimizing inventory management, and deflect potential issues more proactively, for example, by transferring stock to the right location using optimized routing ahead of time.

- **Boost Efficiency with Automated Product Quality Inspection**: By using computer vision models, you can identify product defects faster with improved consistency and accuracy at an early stage so that customers receive high-quality products in a timely fashion. This reduces the number of customer returns and insurance claims that are filed due to issues in product quality, thus saving costs and time.

All the components of supply chain optimization discussed above need to work together as part of the workflow and therefore require low latency and high throughput in order to meet the goal of delivering an optimal quality product to a customer's doorstep in a timely fashion. Using cloud services to build the workflow provides you with greater elasticity and scalability at an optimized cost. Moreover, with native purpose-built services, you can eliminate the heavy lifting and reduce the time to market.

Summary

In this chapter, we started by understanding HPC fundamentals and its importance in processing massive amounts of data to gain meaningful insights. We then discussed the limitations of running HPC workloads on-premises, as different types of HPC applications will have different hardware and software requirements, which becomes time-consuming and costly to procure in-house. Moreover, it will hinder innovation as developers and engineers are limited to the availability of resources instead of the application requirements. Then, we talked about how having HPC workloads on the cloud can help in overcoming these limitations and foster collaboration across global teams, break barriers to innovation, improve architecture design, and optimize performance and cost. Cloud infrastructure has made the specialized hardware needed for HPC applications more accessible, which has led to innovation in this space across a wide range of industries. Therefore, in the last section, we discussed some emerging workloads in HPC, such as in life sciences and healthcare, supply chain optimization, and AVs, along with real-world examples.

In the next chapter, we will dive into data management and transfer, which is the first step to running HPC workloads on the cloud.

Further reading

You can check out the following links for additional information regarding this chapter's topics:

- https://www.statista.com/statistics/871513/worldwide-data-created/
- https://d1.awsstatic.com/whitepapers/Intro_to_HPC_on_AWS.pdf
- https://aws.amazon.com/solutions/case-studies/novartis/
- https://aws.amazon.com/solutions/case-studies/Lyft-level-5-spot/
- https://d1.awsstatic.com/psc-digital/2021/gc-700/supply-chain-ebook/Supply-Chain-eBook.pdf
- https://d1.awsstatic.com/HCLS%20Whitepaper%20.pdf

2
Data Management and Transfer

In *Chapter 1*, *High-Performance Computing Fundamentals*, we introduced the concepts of HPC applications, why we need HPC, and its use cases across different industries. Before we begin developing HPC applications, we need to migrate the required data into the cloud. In this chapter, we will uncover some of the challenges in managing and transferring data to the cloud and the ways to mitigate them. We will dive deeper into the services in **AWS online and offline data transfer services** using which you can securely transfer data to the AWS cloud, while maintaining data integrity and consistency. We will cover different data transfer scenarios and provide guidance on how to select the right service for each one.

We will cover the following topics in this chapter:

- Importance of data management
- Challenges of moving data into the cloud
- How to securely transfer large amounts of data into the cloud
- AWS online data transfer services
- AWS offline data transfer services

These topics will help you understand how you can transfer **Gigabytes (GB)**, **Terabytes (TB)**, or **Petabytes (PB)** of data onto the cloud with minimal disruption, cost, and time involved.

Let's get started with data management and its role in HPC applications.

Importance of data management

Data management is the process of effectively capturing, storing, and collating data created by different applications in your company to make sure it's accurate, consistent, and available when needed. It includes developing policies and procedures for managing your end-to-end data life cycle. The following are some of the elements of the data life cycle specific to HPC applications, due to which it's important to have data management policies in place:

- Cleaning and transforming raw data to perform detailed faultless analysis.

- Designing and building data pipelines to automatically transfer data from one system to another.

- **Extracting, Transforming, and Loading** (ETL) data into appropriate data storage systems such as databases, data warehouses, and object storage or filesystems from disparate data sources.

- Building data catalogs for storing metadata to make it easier to find and track the data lineage.

- Following policies and procedures as outlined by your data governance model. This also involves conforming to the compliance requirements of the federal and regional authorities of the country where data is being captured and stored. For example, if you are a healthcare organization in California, United States, you would need to follow both federal and state data privacy laws, including the **Health Insurance Portability and Accountability Act** (**HIPAA**) and California's health data privacy law, the **Confidentiality of Medical Information Act** (**CMIA**). Additionally, you would also need to follow the **California Consumer Privacy Act** (**CCPA**), which came into effect starting January 1, 2020, as it relates to healthcare data. If you are in Europe, you would have to follow the data guidelines governed by the European Union's **General Data Protection Regulation** (**GDPR**).

- Protecting your data from unauthorized access, while at rest or in transit.

Now that we have understood the significance of data management in HPC applications, let's see some of the challenges of transferring large amounts of data into the cloud.

Challenges of moving data into the cloud

In order to start building HPC applications on the cloud, you need to have data on the cloud, and also think about the various elements of your data life cycle in order to be able to manage data effectively. One way is to write custom code for transferring data, which will be time-consuming and might involve the following challenges:

- Preserving the permissions and metadata of files.

- Making sure that data transfer does not impact other existing applications in terms of performance, availability, and scalability, especially in the case of online data transfer (transferring data over the network).

- Scheduling data transfer for non-business hours to ensure other applications are not impeded.

- In terms of structured data, you might have to think about schema conversion and database migration.

- Maintaining data integrity and validating the transfer.

- Monitoring the status of the data transfer, having the ability to look up the history of previous transfers, and having a retry mechanism in place to ensure successful transfers.

- Making sure there are no duplicates – once the data has been transferred, the system should not trigger the transfer again.

- Protecting data during the transfer, which will include encrypting data both in transit and at rest.

- Ensuring data arrives intact and is not corrupted. You would need a mechanism to check that the data arriving at the destination matches the data read from the source to validate data consistency.

- Last but not least, you would have to manage, version-control, and optimize your data-copying scripts.

The data transfer and migration services offered by AWS can assist you in securely transferring data to the cloud without you having to write and manage code, helping you overcome these aforementioned challenges. In order to select the right service based on your business requirement, you first need to build a data transfer strategy. We will discuss the AWS data transfer services in a subsequent section of this chapter. Let's first understand the items that you need to consider while building your strategy.

In a nutshell, your data transfer strategy needs to take the following into account in order to move data with minimal disruption, time, and cost:

- What kind of data do you need for developing your HPC application – for example, structured data, unstructured data (such as images and PDF documents), or a combination of both?

- For unstructured data, which filesystem do you use for storing your files currently? Is it on **Network Attached Storage (NAS)** or **Storage Area Network (SAN)**?

- How much purchased storage is available right now, and for how long will it last based on the rate of growth of your data, before you plan to buy more storage?

- For structured data, which database do you use?

- Are you tied up in database licenses? If yes, when are they due for renewal, and what are the costs of the licenses?

- What is the volume of data that you need to transfer to the cloud?

- What are the other applications that are using this data?

- Do these applications require local access to data? Will there be any performance impact on the existing applications if the data is moved to the cloud?

- What is your network bandwidth? Is it good enough to transfer data over the network?

- How quickly do you need to move your data to the cloud?

Based on the answers to these questions, you can create your data strategy and select appropriate AWS services that will help you to transfer data with ease and mitigate the challenges mentioned in the preceding list. To understand it better, let's move to the next topic and see how to securely transfer large amounts of data into the cloud with a simple example.

How to securely transfer large amounts of data into the cloud

To understand this topic, let's start with a simple example where you want to build and train a computer vision deep learning model to detect product defects in your manufacturing production line. You have cameras installed on each production line, which capture hundreds of images each day. Each image can be up to 5 MB in size, and you have about 1 TB of data, which is currently stored on-premises in a NAS filesystem that you want to use to train your machine learning model. You have about 1 Gbps of network bandwidth and need to start training your model in 2-4 weeks. There is no impact on other applications if the data is moved to the cloud and no structured data is needed for building the computer vision model. Let's rearrange this information into the following structure, which will become part of your data strategy document:

- **Objective**: To transfer 1 TB of image data to the cloud, where the file size can be up to 5 MB. Need to automate the data transfer to copy about 10 GB of images every night to the cloud. Additionally, need to preserve the metadata and file permissions while copying data to the cloud.

- **Timeline**: 2-4 weeks

- **Data type**: Unstructured data – JPG or PNG format image files

- **Dependency**: None

- **Impact on existing applications**: None

- **Network bandwidth**: 1 Gbps

- **Existing storage type**: Network attached storage

- **Purpose of data transfer**: To perform distributed training on a computer vision deep learning model using multiple GPUs

- **Data destination**: Amazon S3, which is secure, durable, and the most cost-effective object storage on AWS for storing large amounts of data

- **Sensitive data**: None, but data should not be available for public access

- **Local data access**: Not required

Since you have 5 TB of data with a maximum file size of 5 MB to transfer securely to Amazon S3, you can use the AWS DataSync service. It is an AWS online data transfer service to migrate data securely using a **Virtual Private Cloud (VPC)** endpoint to avoid your data going through the open internet. We will discuss all the AWS data transfer services in detail in the later sections of this chapter.

The following architecture visually depicts how the transfer will take place:

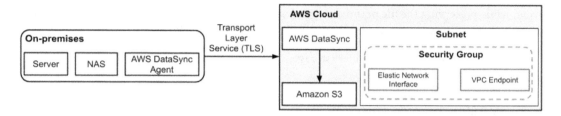

Figure 2.1 – Data transfer using AWS DataSync with a VPC endpoint

The AWS DataSync agent transfers the data between your local storage, NAS in this case, and AWS. You deploy the agent in a **Virtual Machine (VM)** in your on-premises network, where your data source resides. With this approach, you can minimize the network overhead while transferring data using the **Network File System (NFS)** and **Server Message Block (SMB)** protocols.

Let's take a deeper look into AWS DataSync in the next section.

AWS online data transfer services

Online data transfer services are out-of-the-box solutions built by AWS for transferring data between on-premises systems and the AWS cloud via the internet. They include the following services:

- AWS DataSync
- AWS Transfer Family
- Amazon S3 Transfer Acceleration
- Amazon Kinesis
- AWS Snowcone

Let's look at each of these services in detail to understand the scenarios in which we can use the relevant services.

AWS DataSync

AWS DataSync helps you overcome the challenges of transferring data from on-premises to AWS storage services and between AWS storage services in a fast and secure fashion. It also enables you to automate or schedule the data transfer to optimize your use of network bandwidth, which might be shared with other applications. You can monitor the data transfer task, add data integrity checks to make sure that the data transfer was successful, and validate that data was not corrupted during the transfer, while preserving the file permissions and associated metadata. DataSync offers integration with multiple filesystems and enables you to transfer data between the following resources:

- On-premises file servers and object storage:

  - NFS file servers

  - SMB file servers

  - **Hadoop Distributed File System (HDFS)**

  - Self-managed object storage

- AWS storage services:

  - **Snow Family Devices**

  - **Amazon Simple Storage Service (Amazon S3)** buckets

  - **Amazon Elastic File System (EFS)**

  - **Amazon FSx for Windows File Server**

  - **Amazon FSx for Lustre** filesystems

> **Important note**
> We will discuss AWS storage services in detail in *Chapter 4, Data Storage*.

Use cases

As discussed, AWS DataSync is used for transferring data to the cloud over the network. Let's now see some of the specific use cases for which you can use DataSync:

- Hybrid cloud workloads, where data is generated by on-premises applications and needs to be moved to and from the AWS cloud for processing. This can include HPC applications in healthcare, manufacturing, life sciences, big data analytics in financial services, and research purposes.

- Migrate data rapidly over the network into AWS storage services such as Amazon S3, where you need to make sure that data arrives securely and completely. DataSync has encryption and data integrity during transfer enabled by default. You can also choose to enable additional data verification checks to compare the source and destination data.

- Data archiving, where you want to archive the infrequently accessed data (cold data) directly into durable and long-term storage in the AWS cloud, such as **Amazon S3 Glacier** or **S3 Glacier Deep Archive**. This helps you to free up your on-premises storage capacity and reduce costs.

- Scheduling a data transfer job to automatically start on a recurring basis at a particular time of the day to optimize network bandwidth usage, which might be shared with other applications. For example, in the life sciences domain, you may want to upload genomic data generated by on-premises applications for processing and training machine learning models on a daily basis. You can both schedule data transfer tasks and monitor them as required using DataSync.

Workings of AWS DataSync

We will use an architecture diagram to show how DataSync can transfer data between on-premises self-managed storage systems to AWS storage services and between AWS storage resources.

We will start with on-premises storage to AWS storage services.

Data transfer from on-premises to AWS storage services

The architecture in *Figure 2.2* depicts the data transfer from on-premises to AWS storage resources:

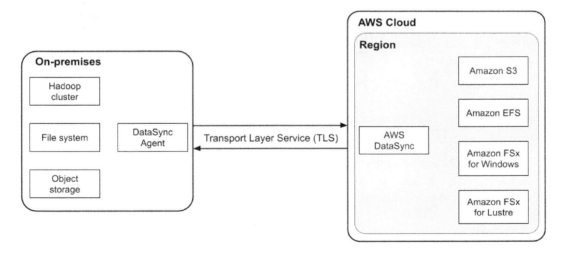

Figure 2.2 – Data transfer from on-premises to AWS storage services using AWS DataSync

The **DataSync agent** is a VM that reads from and writes the data to the on-premises storage. You can configure and activate your agent using the DataSync console or API. This process associates your agent with your AWS account. Once the agent is activated, you can create the data transfer task from the console or API to kick start the data transfer. DataSync encrypts and performs a data integrity check during transfer to make sure that data is transferred securely. You can enable additional checks as well to verify the data copied to the destination is the same as that read at the source. Additionally, you can also monitor your data transfer task. The time that DataSync takes to transfer depends on your network bandwidth, the amount of data, and the network traffic. However, a single data transfer task is capable of utilizing a 10-Gbps network link.

Data transfer between AWS storage resources

Let's take a deeper look to understand the data transfer between AWS storage resources using AWS DataSync.

The architecture in *Figure 2.3* depicts the data transfer between AWS storage resources using DataSync in the same AWS account. The same architecture applies for data transfers within the same region as well as cross-region:

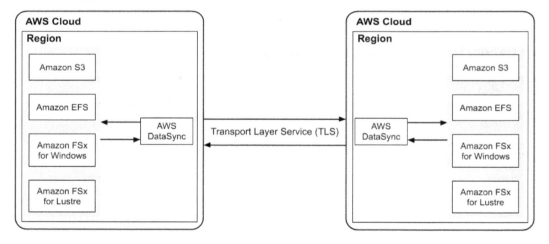

Figure 2.3 – Data transfer between AWS storage resources using AWS DataSync

As shown in the architecture, DataSync does not use the agent for transferring data between AWS resources in the same account. However, if you want to transfer data between different AWS accounts, then you need to set up and activate the DataSync Amazon EC2 agent in an AWS Region.

In summary, you can use AWS DataSync for online data transfer from on-premises to AWS storage services, and between AWS storage resources. AWS DataSync transfers data quickly, safely, and in a cost-effective manner while ensuring data integrity and consistency, without the need to write and manage data-copy scripts.

Now, let's move on to another AWS data transfer service, AWS Transfer Family, which is used for scaling your recurring business-to-business file transfers to Amazon S3 and Amazon EFS.

AWS Transfer Family

File transfer protocols such as **File Transfer Protocol (FTP)**, **Secure File Transfer Protocol (SFTP)**, and **File Transfer Protocol Secure** (**FTPS**) are commonly used in business-to-business data exchange workflows across different industries including financial services, healthcare, manufacturing, and retail. AWS Transfer Family helps in scaling and migrating these file workflows to the AWS cloud. It uses the FTP, SFTP, and FTPS protocols for data transfer. It enables you to transfer files to and from Amazon EFS and Amazon S3.

Use cases

As discussed, AWS Transfer Family uses protocols such as FTP, SFTP, and FTPS for data exchange workflows in business-to-business contexts. So, let's understand some of the common use cases for transferring data to and from Amazon S3 and Amazon EFS using AWS Transfer Family:

- Securely transfer files internally within your organization or with third-party vendors. Some industries, including financial services, life sciences, and healthcare, have to make sure to have a secure file transfer workflow in place due to the sensitive nature of the data and to comply with regulations such as **Payment Card Industry Data Security Standard** (**PCI DSS**), HIPPA, or GDPR, depending on their location.

- To distribute subscription-based content to your customers. For example, BluTV is a famous subscription-based video-on-demand service in Turkey, which is available globally and caters to Turkish and Arabic speaking viewers. Previously, they self-managed their SFTP setup on the cloud, and ran into a lot of problems such as managing open source projects for mounting S3 to Amazon EC2 and scaling issues when additional resources were required. After moving their setup to the fully managed AWS Transfer Family for SFTP, they no longer have to monitor their file transfers, manage open source projects, or pay for unused resources.

- For building a central repository of data (also known as a data lake) on AWS for storing both structured and unstructured data coming from disparate data sources and third parties such as vendors or partners. For example, FINRA, a government-authorized non-profit organization that oversees US stockbrokers, has a data lake on Amazon S3 as its central source of data. FINRA uses the AWS Transfer Family for SFTP service to alleviate operational overheads while maintaining a connection to their existing authentication systems for external users to avoid any disruption while migrating their SFTP services to AWS.

Now that we have gone over some of the use cases for AWS Transfer Family, let's see how it works.

Workings of AWS Transfer Family

The architecture in *Figure 2.4* shows how files are transferred using AWS Transfer Family from on-premises file servers to Amazon S3 or Amazon EFS, which can then be used for downstream file processing workflows such as content distribution, machine learning, and data analysis:

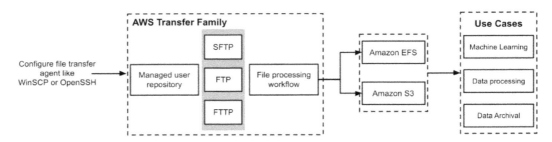

Figure 2.4 – File transfer workflow using AWS Transfer Family

You can configure any standard file transfer protocol client such as WinSCP, FileZilla, or OpenSSH to initially transfer to Amazon S3 or EFS using AWS Transfer Family. It will first authenticate the user based on the identity provider type that you have configured, and once the user is authenticated, it will initiate the file transfer.

So far, we have seen how we can transfer data using AWS DataSync and AWS Transfer Family over the network and understood their use cases and how these services work to transfer data securely while reducing the operational burden in a cost-effective manner. Let's now see how we can accelerate the data transfer to S3 using Amazon S3 Transfer Acceleration.

Amazon S3 Transfer Acceleration

Amazon S3 Transfer Acceleration (**S3TA**) is a feature in Amazon S3 buckets that lets you speed up your data transfer to a S3 bucket over long distances, regardless of internet traffic and without the need for any special clients or proprietary network protocols. You can speed up transfers to and from Amazon S3 by 50-500% using the transfer acceleration feature.

Some of the use cases include the following:

- Time-sensitive transfers of large files, such as lab imagery or media, from distributed locations to your data lake built on Amazon S3 (centralized data repository).

- Web or mobile applications with a file upload or download feature where users are geographically distributed and are far from the destination S3 bucket. S3TA can accelerate this long-distance transfer of files and helps you provide a better user experience.

It uses Amazon CloudFront's globally distributed edge locations, AWS backbone networks, and network protocol optimizations to route traffic, which speeds up the transfer, reduces the internet traffic variability, and helps in logically shortening the distance to S3 for remote applications.

> **Important note**
> There is an additional charge to use Amazon S3 Transfer Acceleration.

We have discussed using online data transfer services such as AWS DataSync, AWS Transfer Family, and S3TA for moving data from on-premises storage to AWS storage resources over the network. There might be scenarios where you want to transfer streaming data in real time to the AWS cloud, for example, telemetry data from IoT sensors, video for online streaming applications, and so on. For this, we will go deeper into Amazon Kinesis, which is a fully managed streaming service built by AWS.

Amazon Kinesis

Amazon Kinesis is a **fully managed** service used to collect, process, and analyze streaming data in **real time at any scale**. Streaming data can include ingesting application logs, audio, video, website clickstreams, or IoT sensor data for deep learning, machine learning, analytics, and other applications. It allows you to perform data analysis as the data arrives in real time, instead of waiting for all the data to be transferred before processing.

Amazon Kinesis includes the following services:

- **Kinesis Video Streams**: This is used when you have to securely stream video from connected devices to AWS for applications such as processing, analytics, or machine learning to drive insights in real time. It has a built-in autoscaling mechanism to provision the infrastructure required for ingesting video streams coming from millions of devices. It automatically encrypts the data at rest as well as in transit. It uses Amazon S3 as its underlying storage, allowing you to store and retrieve data reliably. This helps you to develop real-time computer vision applications by integrating it with other fully managed AWS services or using popular open source machine learning frameworks on AWS. *Figure 2.5* shows how Kinesis video streams can be used to collect, process, and store video streams coming from media devices for machine learning, analytics, and playback with integration with other media applications:

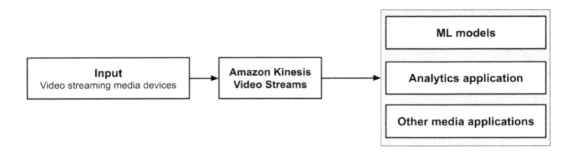

Figure 2.5 – Capture, process, and store video streams for machine learning, analytics, and playback

- **Kinesis Data Streams**: This is a fully managed serverless service used for securely streaming data at any scale in a cost-effective manner. You can stream gigabytes of data per second by adjusting your capacity or use it in on-demand mode for automatic scaling and provisioning the underlying infrastructure based on the capacity required by your application. It has built-in integration with other AWS services, and you only pay for what you use. *Figure 2.6* shows how Kinesis Data Streams can be used to ingest, process, and store streaming data generated by different sources to other AWS services to gain insights in real time:

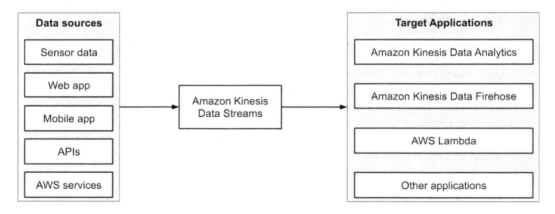

Figure 2.6 – Capture data from different sources into Amazon Kinesis Data Streams

- **Kinesis Data Firehose**: This is used to securely stream data into data lakes built on Amazon S3, or data warehouses such as Amazon Redshift, into the required formats for further processing or analysis without the need to build data processing pipelines. Let's look at some of the benefits of using Kinesis Data Firehose:

  - It enables you to create your delivery stream easily by extracting, transforming, and loading streaming data securely at scale, without the need to manage the underlying infrastructure.

- It has built-in autoscaling to provision the resources required by your streaming application, without any continuous management.

- It enables you to transform raw streaming data using either built-in or custom transformations. It supports converting data into different formats such as Apache Parquet and can dynamically partition data without building any custom processing logic.

- You can enhance your data streams using machine learning models within Kinesis Firehose to analyze and perform inference as the data travels to the destination. It provides enhanced network security by monitoring and creating alerts in real time when there is a potential threat using **Security Information and Event Management (SIEM)** tools.

- You can connect with 30+ AWS services and streaming destinations, which are fully integrated with Kinesis Data Firehose.

Figure 2.7 shows how Amazon Kinesis Data Firehose can be used for ETL use cases without having to write long lines of code or managing your own infrastructure at scale:

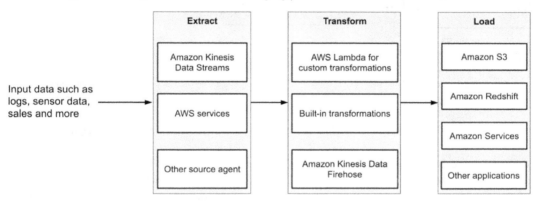

Figure 2.7 – ETL using Amazon Kinesis Data Firehose

- **Kinesis Data Analytics**: This is used to process data streams in real time with serverless and fully managed **Apache Flink** or SQL, from data sources such as Amazon S3, Amazon Kinesis Data Streams, and **Amazon Managed Apache Kafka (MSK)** (used to ingest and process streaming data). It can also be used to trigger real-time actions such as anomaly detection from long-running stateful computations based on past data trends. It has a built-in autoscaling mechanism to match the volume and throughput of your input data stream. You only pay for what you use; there is no minimum fee or set-up cost associated with it. It helps you to understand your data in real time, for example, by building a leaderboard for your gaming application, analyzing sensor data, log analytics, web clickstream analytics, building a streaming ETL application, or continuously generating metrics for understanding data trends.

Figure 2.8 shows how a typical Kinesis Data Analytics application works. It has three main components:

- An input data streaming source on which to perform real-time analytics

- **Amazon Kinesis Data Analytics Studio Notebook** to analyze streaming data using SQL queries and Python or Scala programs

- Finally, it stores processed results on a destination service or application, such as Amazon Redshift or Amazon DynamoDB (NoSQL database) and Amazon Kinesis Data Streams:

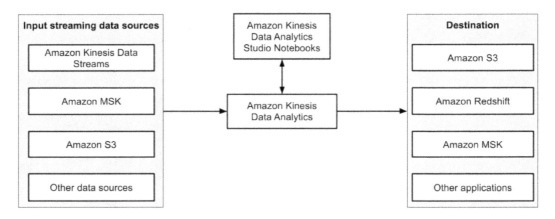

Figure 2.8 – Real-time processing of streaming data using Amazon Kinesis Data Analytics

In this section, we discussed how to transfer and process streaming data to AWS Storage, using Amazon Kinesis. There are some use cases, such as edge computing and edge storage, for which you can use **AWS Snowcone**, a portable, rugged, and secure device for edge computing, storage, and data transfer. Next, let's see how we can transfer data online from AWS Snowcone to AWS.

AWS Snowcone

AWS Snowcone is a small, rugged, and portable device, used for running edge computing workloads, edge storage, and data transfer. The device weighs about 4.5 lbs (2.1 kg) and has multiple layers of security and encryption. It has 8 TB of storage, while the AWS Snowcone **Solid State Drive** (**SSD**) version provides 14 TB. Some of the common use cases for Snowcone are as follows:

- Healthcare IoT, for transferring critical and sensitive data from emergency medical vehicles to hospitals for processing data faster and reducing the response time to serve the patients in a better way. You can then transfer the data securely to the AWS cloud.

- Industrial IoT, for capturing sensor or machine data, as it can withstand extreme temperatures, vibrations, and humidity found on factory floors where traditional edge devices might not work.

- Capturing and storing sensor data from autonomous vehicles and drones.

You can transfer terabytes of data from various AWS Snowcone devices over the network using AWS DataSync, as discussed in the *AWS DataSync* section.

AWS online data transfer services are helpful when you have to transfer up to terabytes of data over the network to AWS. The time taken to transfer data is dependent on your available network bandwidth and internet traffic. When you have to transfer data from remote locations, or when your network bandwidth is heavily used by existing applications, you would need an alternative mechanism to transfer data offline. Let's discuss the AWS offline data transfer services in the next section.

AWS offline data transfer services

For transferring up to petabytes of data via offline methods, in a secure and cost-effective fashion, you can use **AWS Snow Family devices**. Sometimes, your applications may require enhanced performance at the edge, where you want to process and analyze your data close to the source in order to deliver real-time meaningful insights. This would mean having AWS-managed hardware and software services beyond the AWS cloud. AWS Snow Family can help you to run operations outside of your data center, as well as in remote locations with limited network connectivity.

It consists of the following devices:

- **AWS Snowcone**: In the *AWS online data transfer services* section, we introduced and discussed how Snowcone can be used for collecting and storing data at the edge and then transferring it to the AWS cloud using AWS DataSync. In cases of limited network bandwidth, you can also use it for offline data transfer by sending the device to an AWS facility. It includes an E Ink shipping label, which also aids in tracking.

- **AWS Snowball**: This is used for migrating data and performing edge computing. It comes with two options:

  - Snowball Edge Compute Optimized has 42 TB of block or Amazon S3-compatible storage with 52 vCPUs and an optional GPU for edge computing use cases such as machine learning, video analysis, and big data processing in environments with intermittent network connectivity, such as industrial and transportation use cases, or extremely remote locations found in defense or military applications.

  - Snowball Edge Storage Optimized has 80 TB of usable block or Amazon S3-compatible object storage, with 40 vCPUs for performing computing at the edge. It is primarily used for either local storage or large-scale offline data transfers to the AWS cloud.

- AWS Snowmobile is used to migrate up to 100 PB of data in a 45-foot-long shipping container, which is tamper resistant, waterproof, and temperature controlled with multiple layers of logical and physical security. It is ideal for use cases where you have to transfer exabytes or hundreds of petabytes of data, which might occur due to data center shutdowns. You need to order it from the AWS Snow Family console, and it arrives at your site as a network-attached data store that connects to your local network to perform high-speed data transfers. Once the data is moved to the device, it is driven back to the AWS facility where the data is then uploaded to the specified Amazon S3 bucket. To ensure data security in transit and the successful delivery of data to the AWS facility, it comes with fire suppression, encryption, dedicated security personnel, GPS tracking, alarm monitoring, 24/7 video surveillance, and an escort security vehicle.

Now that we understand the various offline data transfer options offered by AWS, let's understand the process for ordering the device.

Process for ordering a device from AWS Snow Family

To order AWS Snowmobile, you need to contact AWS sales support. For Snowcone or Snowball devices, you can follow these steps:

1. Log in to the AWS Console and type AWS Snow Family in the search bar. Click on **AWS Snow Family**, which will take you to the AWS Snow Family console as shown in *Figure 2.9*:

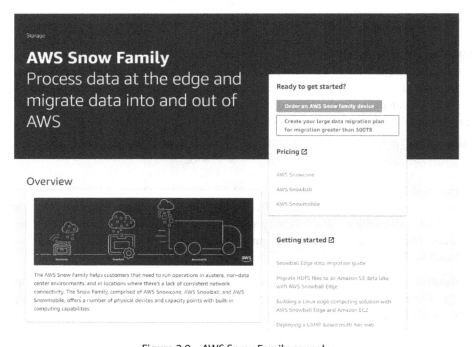

Figure 2.9 – AWS Snow Family console

2. Click on the orange button reading **Order an AWS Snow Family device**, which opens another screen, as shown in *Figure 2.10*. This will provide you with steps to create a job for ordering the device:

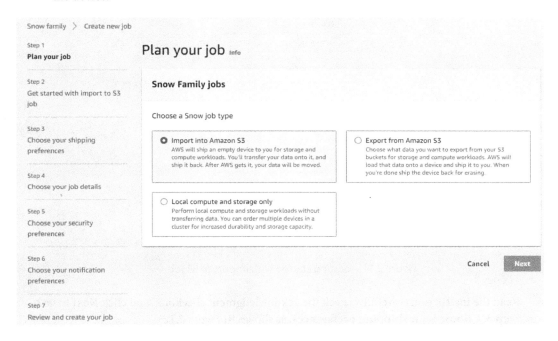

Figure 2.10 – AWS Snow Family – Create new job

3. Click on **Next** to go to **Step 2**, **Get started with import to S3 job**, as shown in *Figure 2.11*:

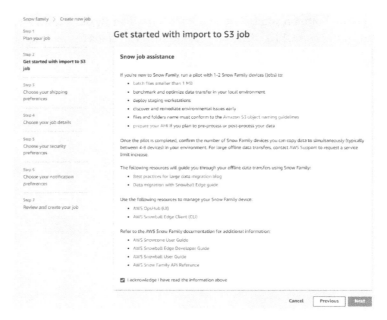

Figure 2.11 – Getting started with import to S3 job

4. Read the instructions carefully, check the acknowledgment checkbox, and click **Next** to go to **Step 3**, **Choose your shipping preferences**, as shown in *Figure 2.12*:

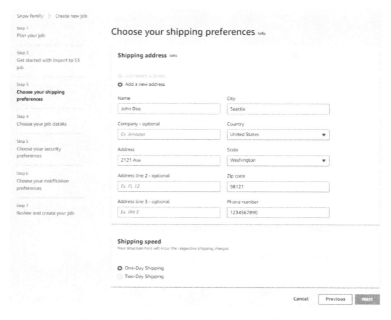

Figure 2.12 – Choose your shipping preferences

5. Fill in your shipping details, select your preferred shipping speed, and click **Next** to go to **Step 4**, **Choose your job details**, as shown in *Figure 2.13*. Make sure to enter a valid address as it will give you an error message if your address is incorrect:

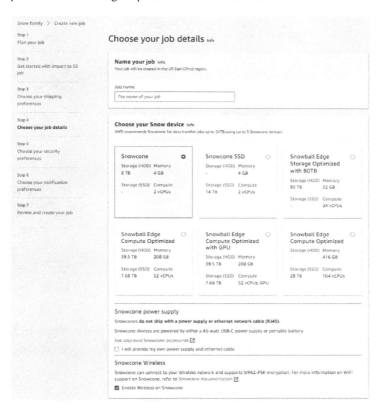

Figure 2.13 – Choose your job details

On this screen, you can select your Snow device, power supply, wireless options for Snowcone, S3 bucket, compute using EC2 instances, and the option to install the AWS IoT Greengrass validated AMI.

Please note that the S3 bucket will appear as directories on your device, and the data in those directories will be transferred back to S3. If you have selected the AWS IoT Greengrass AMI to run IoT workloads on the device, you also need to select the **Remote device management** option to open and manage the device remotely with OpsHub or Snowball Client.

Important note

All the options as mentioned in *step 4* are not shown in *Figure 2.13 – Choose your job details*, but will be present on your console screen.

6.   After you have filled out the job details, click on the **Next** button to go to **Step 5**, **Choose your security preferences**, as shown in *Figure 2.14*:

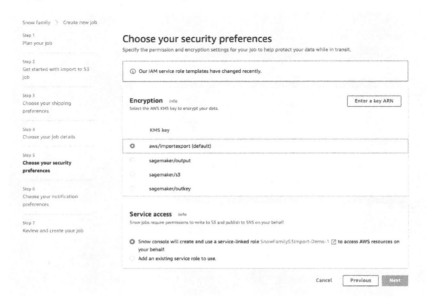

Figure 2.14 – Choose your security preferences

7.   Select the permission and encryption settings for your job on this screen, which will help you to protect your data while in transit, and then click on **Next** to go to **Step 6**, **Choose your notification preferences**, as shown in *Figure 2.15*:

Figure 2.15 – Choose your notification preferences

8. To receive email notifications of your job status changes, you can choose either an existing **Simple Notification Service (SNS)** topic or create a new SNS topic. Click on **Next** to go to **Step 7**, **Review and create your job**, as shown in *Figure 2.16*:

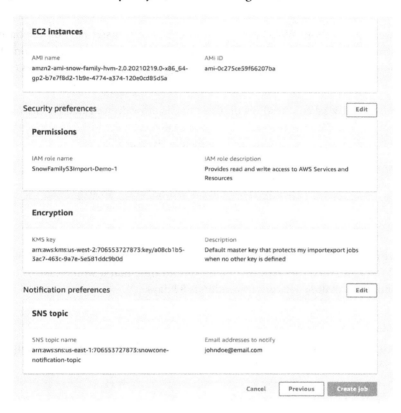

Figure 2.16 – Review and create your job

9. You can review all the details that you have entered from **Step 1** through **Step 7** and then click on the **Create job** button.

10. Once the job is created, it will take you to the **Jobs** screen, as shown in *Figure 2.17*, where you can see details of your job including the status:

Figure 2.17 – Snow Family jobs

> **Important note**
>
> On the **Actions** drop-down menu, you also have options to cancel a job, edit a job name, and clone a job.

In this section, we learned about AWS Snow Family devices to transfer data offline based on our application requirements, network connectivity, available bandwidth, and the location of our data sources. We also discussed how we can use these devices not only for transferring data but also for edge computing.

One of the most frequently asked questions on this topic is, how do we calculate the time taken to move data to the cloud based on the network speed and available bandwidth? For this, AWS provides a simple formula based on the best-case scenario, which is given as follows:

$$\text{Number of days} = (\text{Total Bytes})/ (\text{Megabits per second} * 125 * 1000 * \text{Network Utilization} * 60 \text{ seconds} * 60 \text{ minutes} * 24 \text{ hours})$$

For example, if we have a network connection of 1.544 Mbps, and we want to move 1 TB of data into and out of the AWS cloud, then theoretically the minimum time that it would take to transfer over your network connection at 80% network utilization is 82 days.

> **Important note**
>
> Please note that this formula only gives a high-level estimate; the actual time taken might differ based on the variability of network traffic and available bandwidth.

Let's now take a brief look at all the topics that we have covered in this chapter.

Summary

In this chapter, we talked about various aspects of data management, including data governance and compliance with the legal requirements of federal and regional authorities of the country where the data resides. We also discussed that in order to build HPC applications on the cloud, we need to have data on the cloud, and looked at the challenges of transferring this data to the cloud. In order to mitigate these challenges, we can use the managed AWS data transfer services, and in order to select which service to use for your application, we then discussed the elements of building a data strategy.

We then took an example of how we can transfer petabyte-scale data to the cloud in order to understand the concepts involved in a data transfer strategy. Finally, we did a deep dive on various AWS data transfer services for both online and offline data transfer based on your network bandwidth, connectivity, type of application, speed of data transfer, and location of your data source.

Now that we understand the mechanisms for transferring data to the cloud, the challenges involved, and how to mitigate them, in the next chapter, we will focus on understanding the various compute options provided by AWS for running HPC applications, and how to optimize these based on the application requirements.

Further reading

The following are some additional resources for this chapter:

- *Migrate Petabyte Scale Data*: https://aws.amazon.com/getting-started/projects/migrate-petabyte-scale-data/

- *Introduction to HPC on AWS*: https://d1.awsstatic.com/whitepapers/Intro_to_HPC_on_AWS.pdf

- *Data management versus data governance*: https://www.tableau.com/learn/articles/data-management-vs-data-governance

- *What is DataSync?*: https://docs.aws.amazon.com/datasync/latest/userguide/what-is-datasync.html

- *S3 Transfer Acceleration*: https://aws.amazon.com/s3/transfer-acceleration/

- *AWS Transfer Family Customers*: https://aws.amazon.com/aws-transfer-family/customers/

- *Kinesis Data Streams*: https://aws.amazon.com/kinesis/data-streams/

- *Kinesis Video Streams*: https://aws.amazon.com/kinesis/video-streams/

- *Kinesis Data Analytics*: https://aws.amazon.com/kinesis/data-analytics/

- *AWS Snowcone*: https://aws.amazon.com/snowcone

- *AWS Snow Family*: https://aws.amazon.com/snow/

- *Cloud Data Migration*: https://aws.amazon.com/cloud-data-migration/

3

Compute and Networking

Several large and small organizations run workloads on AWS using AWS compute. Here, AWS **Compute** refers to a set of services on AWS that help you build and deploy your own solutions and services; this can include workloads as diverse as websites, data analytics engines, **Machine Learning** (**ML**), **High-Performance Computing** (**HPC**), and more. Being one of the first services to be released, **Amazon Elastic Compute Cloud** (**EC2**) is sometimes used synonymously with the term *compute* and offers a wide variety of instance types, processors, memory, and storage configurations for your workloads.

Apart from EC2, compute services that are suited to some specific types of workloads include Amazon **Elastic Container Service** (**ECS**), **Elastic Kubernetes Service** (**EKS**), Batch, Lambda, Wavelength, and Outposts. **Networking on AWS** refers to foundational networking services, including Amazon **Virtual Private Cloud** (**VPC**), AWS Transit Gateway, and AWS PrivateLink. These services, along with the various compute services, enable you to build solutions with the most secure and performant networked systems at a global scale. AWS compute and networking concepts are two broad topics and are important to understand many concepts that will be discussed in the following chapters.

Compute and networking also form two important pillars of HPC, along with data management, which was discussed in the last chapter. Every application of HPC is generally optimized for high levels of distributed compute, which depends on networking.

In this chapter, you will learn about the different services AWS offers for compute and networking, how these services are used for different types of computing workloads, and lastly, best practices for the HPC type of workloads on AWS, which goes beyond the AWS Well-Architected Framework.

Specifically, in this chapter, we will cover the following topics:

- Introducing the AWS compute ecosystem
- Networking on AWS
- Selecting the right compute for HPC workloads
- Best practices for HPC workloads

Introducing the AWS compute ecosystem

Compute lies at the foundation of every HPC application that you will read about in and outside of this book. In AWS and other clouds in general, compute refers to a group of services that offer the basic building blocks of performing a computation or some business logic. This can range from basic data computations to ML.

The basic units of measuring compute power on AWS (regardless of the service we are talking about) are as follows:

- *Processing units* – this can be measured as the number of **Central Processing Units** (**CPUs**), **Virtual CPUs** (**vCPUs**), or **Graphics Processing Units** (**GPUs**)
- *Memory* – this is the total requested or allocated memory for the application measured in units of bytes

Typical HPC applications access multiple instances and hence can take advantage of pooled compute and memory resources for larger workloads.

The foundational service that provides compute resources for customers to build their applications on AWS is called Amazon EC2. Amazon EC2 provides customers with a choice of about 500 instance types (at the time of writing this book and according to public documentation). Customers can then tailor the right combination of instance types for their business applications.

Amazon EC2 provides five types of instances:

- General purpose instances
- Compute optimized instances
- Accelerated computing instances
- Memory optimized instances
- Storage optimized instances

Each of the instance types listed here is actually a family of instances, as shown in *Figure 3.1*:

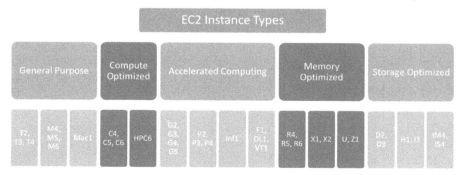

Figure 3.1 – Amazon EC2 instance types

In the following section, we will highlight some important facts about these instance types.

General purpose instances

General purpose instances can be used for a variety of workloads. They have the right balance of compute, memory, and storage for most typical applications that customers have on AWS. On AWS, there are several types of general purpose instances:

- **T-type instances**: T instances, for example, the T2 instance, are *burstable* instances that provide a basic level of compute for low compute- and memory-footprint workloads. The following figure shows what a typical workload that is suited for T-type instances might look like – most of the time is spent under the baseline CPU utilization level, with a need to *burst* above this baseline occasionally. With non-burstable instances, application owners need to over-provision for the burst CPU levels, and therefore pay more while utilizing very little. With burstable T instances, credits are accrued for utilization under the baseline (white areas below the baseline), and these credits can be used when the application is experiencing higher loads; see the gray filled-in areas in the following graph:

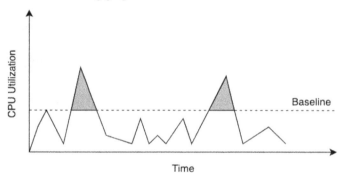

Figure 3.2 – CPU utilization versus time for burstable T-type instances on AWS

- **M-type instances**: M instances (like the M4, M5, and M6) can be used for a variety of workloads that need a balance of compute, memory, and networking, including (but not limited to) web and application servers, and small to medium-sized database workloads. Special versions of M5 and M6 instances are offered that are suitable for certain workloads. For example, the M5zn instance types can be used for applications that demand extremely high single-threaded performance, such as HPC, simulations, and gaming.

- **A-type instances**: A1 instances are used to run **Advanced RISC Machine** or **ARM**-based applications, such as microservices and web servers powered by the AWS Graviton ARM processors.

- **Mac-type instances**: Mac1 instances are powered by Apple's Mac Mini computers and provide very high network and storage bandwidth. They are typically used for building and testing Apple applications for the iPhone, Mac, and so on.

In the following section, we will discuss compute optimized instances on AWS.

Compute optimized instances

Many HPC applications that will be described in this book take advantage of the high-performance, compute optimized instance types on AWS.

There are several types of compute optimized instances:

- **C5 instances**: C5 and C5n instances provide low-cost, high-performance compute for typical HPC, gaming, batch processing, and modeling. C5 instances use Intel Xeon processors (first and second generation) and provide upward of 3.4 GHz clock speeds on a single core. C5a instances also provide AMD processors for high performance at an even lower cost. C5n instances are well suited to HPC applications since they support **Elastic Fabric Adapter** (**EFA**) and can deliver up to 100 gigabits per second of networking throughput. For more information about EFA, please visit https://aws.amazon.com/hpc/efa/.

- **C6 instances**: C6g instances are ARM-based instances based on the AWS Graviton processor. They are ideal for running HPC workloads, ad serving, and game servers. C6g instances are available with local **Non-Volatile Memory express** or **NVMe**-based high performance, low latency SSD storage with 100 gigabits per second networking, and support for EFA. On the other hand, the C6i class of instances is Intel Xeon-based and can provide up to 128 vCPUs per instance for typical HPC workloads.

- **HPC instances**: The HPC6a instance type is powered by third-generation AMD processors for lower cost-to-performance ratios for typical HPC workloads. These instance types also provide 96 CPU cores and 384 GB of RAM for memory-intensive applications. HPC6a instances also support EFA-based networking for up to 100 gigabits per second of throughput.

In the following section, we will discuss accelerated compute instances on AWS.

Accelerated compute instances

Accelerated computing instances use co-processors such as GPUs to accelerate performance for workloads such as floating point number calculations useful for ML, deep learning, and graphics processing.

Accelerated compute instances use hardware-based compute accelerators such as the following:

- GPUs
- **Field Programmable Gate Arrays** (**FPGAs**)
- AWS Inferentia

GPUs

GPUs were originally used for 3D graphics but are now being used as general-purpose co-processors for various applications such as HPC and deep learning. HPC applications are computation and bandwidth-heavy. Several types of NVIDIA GPUs are available on AWS, and detailed information can be found at the following link, `https://aws.amazon.com/nvidia/`.

Let's dive into the basics of how GPUs help with compute-heavy calculations. Imagine adding a list of numbers to another list of the same size. Visually, this looks like the following diagram:

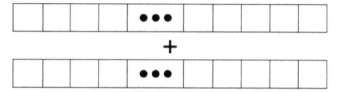

Figure 2.3 – Adding two arrays

The naïve way of adding these to arrays is to loop through all elements of each array and add each corresponding number from the top and bottom arrays. This may be fine for small arrays, but what about arrays that are millions of elements long? To do this on a GPU, we first allocate memory for these two very long arrays and then use *threads* to parallelize these computations. Adding these arrays using a single thread on a single GPU is the same as our earlier naïve approach. Using multiple threads (say 256) can help parallelize this operation by allocating a part of the work to each thread. For example, the first few elements (the total size divided by 256 in this case) will be done by the first thread, and so on. This speeds up the operation by letting each thread focus on a smaller portion of the work and do each of these split-up addition operations in parallel; see the shaded region in the following diagram:

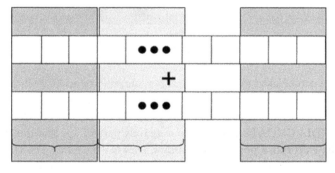

Figure 3.4 – Multiple threads handling a portion of the computation

GPUs today are architected in a way that allows even higher levels of parallelism – multiple processing threads make up a block, and there are usually multiple blocks in a GPU. Each block can run concurrently in a **Streaming Multiprocessor** (**SM**) and process the same set of computations or *kernels*. Visually, this looks like the following:

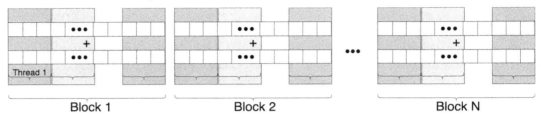

Figure 3.5 – Multiple blocks in a GPU

To give you an idea of what you can access on AWS, consider the **P4d.24xlarge** instance. This instance has eight GPUs, as seen in the following figure, each of which is an NVIDIA A100 housing 108 SMs, with each SM capable of running 2,048 threads in parallel:

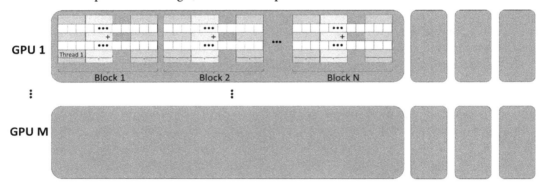

Figure 3.6 – A single instance with multiple GPUs

On AWS, P4d instances can be used to provision a supercomputer or an EC2 Ultracluster with more than 4,000 A100 GPUs, Petabit-scale networking, and scalable, shared high throughput storage on Amazon FSx for Lustre (https://aws.amazon.com/fsx/lustre/). Application and package developers use the NVIDIA CUDA library to build massively parallel applications for HPC and deep learning. For example, PyTorch, a popular ML library, uses NVIDIA's CUDA GPU programming library for training large-scale models. Another example is Ansys Fluent, a popular **Computational Fluid Dynamics** (**CFD**) simulation software that uses GPU cores to accelerate fluid flow computations.

On AWS, there are several families of GPU instances:

- **G family of instances**: G2, G3, G4, and G5 type instances on AWS provide cost-effective access to GPU resources. Each G-type instance mentioned here comes with a different NVIDIA GPU – for example, the latest G5 instances come with NVIDIA A10G GPUs, G5g instances with NVIDIA T4G GPUs, and G4Dn with the NVIDIA Tesla GPUs. AMD-based GPUs are also available – for example, the G4ad instances use the AMD Radeon Pro V520 GPUs.

- **P family of instances**: These instances provide extremely high-performance GPUs for single-instance and distributed applications. The P2 instances provide access to the NVIDIA K80 GPUs, P3 instances have the NVIDIA Tesla V100 GPUs, and the P4d instances have the A10 GPUs.

- **VT1 instances**: These provide access to the Xilinx Alveo U30 media accelerator cards that are primarily used for video transcoding applications. More information about VT1 instances can be found here: https://aws.amazon.com/ec2/instance-types/vt1/.

- **AWS Inferentia**: These instances are specifically designed to provide cost-effective and low latency ML inference capability and are a custom-made chip created by AWS. A typical workflow that customers could follow using these *Inf1* instances is to use an ML framework like TensorFlow to train a model on another EC2 instance or SageMaker training instances, then use Amazon SageMaker's compilation feature *Neo* to compile the model for use with the Inf1 instances. You can also make use of the AWS Neuron SDK to profile and deploy deep learning models onto *Inf1* instances.

FPGA instances

Amazon EC2 F1 instances allow you to develop and deploy hardware-accelerated applications easily on the cloud. Example applications include (but are not limited to) big data analytics, genomics, and simulation-related applications. Developers can use high-level C/C++ code to program their applications, register the FPGA as an **Amazon FPGA Image** (**AFI**), and deploy the application to an F1 instance. For more information on F1 instances, please refer to the links in the Reference section at the end of this chapter.

In the following section, we will discuss memory optimized compute instances on AWS.

Memory optimized instances

Memory optimized instances on AWS are suited to run applications that require storage of extremely large data in memory. Typical applications that fall into this category are in-memory databases, HPC applications, simulation, and **Electronic Design Automation** (**EDA**) applications. On AWS, there are several types of memory optimized instances:

- **R5 instances**: The R5 family of instances (such as the R5, R5a, R5b, and R5n instance types) is a great choice for relational databases such as MySQL, MongoDB, and Cassandra, for in-memory databases such as Redis and Memcached, and business intelligence applications, such as SAP HANA and HPC applications. The R5 metal instance type also provides direct access to processors and memory on the physical server that the instance is based on.

- **R6 instances**: R6 instances such as R6g and R6gd are based on ARM-based AWS Gravitron2 processors and can provide better price-to-performance ratios compared to R5 instances. Application developers can use these instance types to develop or support ARM-based applications that need a high memory footprint.

- **U instances**: These instances are extremely high memory instances – they can offer anywhere from 6 to 24 TB of memory per instance. They are typically used to run large in-memory applications such as databases and SAP HANA and are powered by Intel Xeon Platinum 8176M processors.

- **X instances**: X-type instances (such as X1, X1e, and X1gd instance types) are designed for large-scale in-memory applications in the cloud. Each X1 instance is powered by four Intel Xeon E8880 processors with up to 128 vCPUs, and up to 1,952 GB of memory. X1 instances are picked by developers for their low price-to-performance ratio given the amount of memory provided, compared to other families of instances. X1e instances provide even higher memory (up to 3,904 GB) and support production-grade SAP workloads. Both X1 and X1e instance types provide up to 25 gigabits per second of network bandwidth when used with an **Elastic Network Adapter** (ENA). Finally, X1gd and X2gd are AWS Graviton2-based ARM instances that provide better price performance compared to x86-based X1 instances.

- **Z instance types**: The Z1d instance type provides both high performance and high memory for typical data analytics, financial services, and HPC applications. Z1d instances are well suited for applications that need very high single-threaded performance, with an added dependence on high memory. Z1d instances come in seven different sizes, and up to 48 vCPUs and 384 GB of RAM, so customers can choose the right instance size for their application.

Storage optimized instances

Storage optimized instances are well suited for applications that need frequent, sequential reads and writes from local storage by providing very high **I/O Operations Per Second** (IOPS). There are several storage optimized instances on AWS:

- **D-type instances**: Instances such as D2, D3, and D3en provide high-performance local storage and can be used for MapReduce-style operations (with Hadoop or Spark), log processing, and other big data workloads that do not require all the data to be held in memory but require very fast, on-demand access to this data.

- **H-type instances**: The H1 instance is typically used for MapReduce applications and distributed file storage, and other data-intensive applications.

- **I-type instances**: I type instances such as I3 and I3en are well suited for relational and non-relational databases, in-memory caches, and other big data applications. The I3 instances are NVMe **Solid State Drive** (SSD) -based instances that can provide up to 25 GB of network bandwidth and 14 gigabits per second of dedicated bandwidth to attached **Elastic Block Store** (EBS) volumes. The Im4gn and Is4gen type instances can be used for relational and *NoSQL* databases, streaming applications, and distributed file applications.

Amazon Machine Images (AMIs)

Now that we have discussed different instance types that you can choose for your applications on AWS, we can move on to the topic of **Amazon Machine Images (AMIs)**. AMIs contain all the information needed to launch an instance. This includes the following:

- The description of the operating system to use, the architecture (32 or 64-bit), any applications to be included along with an application server, and EBS snapshots to be attached before launch
- Block-device mapping that defines which volumes to attach to the instance on launch
- Launch permissions that control which AWS accounts can use this AMI

You can create your own AMI, or buy, share, or sell your AMIs on the AWS Marketplace. AWS maintains Amazon Linux-based AMIs that are stable and secure, updated and maintained on a regular basis, and includes several AWS tools and packages. Furthermore, these AMIs are provided free of charge to AWS customers.

Containers on AWS

In the previous section, we spoke about AMIs on AWS that can help isolate and replicate applications across several instances and instance types. Containers can be used to further isolate and launch one or more applications onto instances. The most popular flavor of containers is called Docker. **Docker** is an open platform for developing, shipping, and running applications. Docker provides the ability to package and run an application in a loosely isolated environment called a container. Docker containers are definitions of runnable images, and these images can be run locally on your computer, on virtual machines, or in the cloud. Docker containers can be run on any host operating system, and as such are extremely portable, as long as Docker is running on the host system.

A Docker container contains everything that is needed to run the applications that are defined inside it – this includes configuration information, directory structure, software dependencies, binaries, and packages. This may sound complicated, but it is actually very easy to define a Docker image; this is done in a Dockerfile that may look similar to this:

```
FROM python:3.7-alpine
COPY . /app
WORKDIR /app
RUN pip install -r requirements.txt
CMD ["gunicorn", "-w 4", "main:app"]
```

The preceding file named `Dockerfile` defines the Docker image to run a sample Python application using the popular `Gunicorn` package (see the last line in the file). Before we can run the application, we tell Docker to use the Python-3.7 base image (FROM `python:3.7-alpine`), copy all the required files from the host system to a folder called `app`, and install requirements or dependencies for that application to run successfully (`RUN pip install -r requirements.txt`). Now you can test out this application locally before deploying it at scale on the cloud.

On AWS, you can run containers on EC2 instances of your choice or make use of the many container services available:

- When you need to run containers with server-level control, you can directly run the images that you define on EC2. Furthermore, running these containers on EC2 *Spot* instances can save you up to 90% of the cost over on-demand instances. For more information on *Spot* instances please refer to `https://aws.amazon.com/ec2/spot/`.

- At the opposite end of the spectrum, you can use a service like AWS Fargate to run containers without managing servers. Fargate removes all the operational overhead of maintaining server-level software so you can focus on just the application at hand. With Fargate, you only pay for what you use – for example, if you create an application that downloads data files from Amazon S3, processes these files, and writes output files back to S3, and this process takes 30 minutes to finish, you only pay for the time and resources (vCPUs) used to complete the task.

- When you have multiple, complex, container-based applications, managing and orchestrating these applications is an important task. On AWS, container management and orchestration can be achieved using services like Amazon **Elastic Container Registry** (**ECR**), Amazon **Elastic Container Service** (**ECS**), and Amazon **Elastic Kubernetes Service** (**EKS**). A detailed discussion about these services is outside the scope of this book but if you are interested, you can refer to the links mentioned in the *References* section to learn more.

Serverless compute on AWS

In the previous section, you read about AWS Fargate, which lets you run applications and code based on Docker containers, without the need to manage infrastructure. This is an example of a serverless service on AWS. AWS offers serverless services that have the following features in common:

- No infrastructure to manage
- Automatic scaling
- Built-in high availability
- Pay-per-use billing

Serverless compute technologies on AWS are AWS Lambda and Fargate. AWS Lambda is a serverless computing service that lets you run any code that can be triggered by over 200 services and SaaS applications. Code can be written in popular languages such as Python, `Node.js`, Go, and Java or can be packaged as Docker containers, as described earlier. With AWS Lambda, you only pay for the number of milliseconds that your code runs, beyond a very generous free tier of over a million free requests. AWS Lambda supports the creation of a wide variety of applications including file processing, streaming, web applications, IoT backend applications, and mobile app backends.

For more information on serverless computing on AWS, please refer to the links included in the *References* section.

In the next section, we will cover basic concepts around networking on AWS.

Networking on AWS

Networking on AWS is a vast topic that is out of the scope of this book. However, in order to easily explain some of the sections and chapters that follow, we will attempt to provide a brief overview here. First, AWS has a concept called **regions**, which are physical areas around the world where AWS places clusters of data centers. Each region contains multiple logically separated, groups of data centers called **availability zones**. Each availability zone has independent power, cooling, and physical security. Availability zones are connected via redundant and ultra-low latency AWS Networks. At the time of writing this chapter, AWS has 26 regions and 84 availability zones.

The next foundational concept we will discuss here is a **Virtual Private Cloud** (**VPC**). A VPC is a logical partition that lets you launch and group AWS resources. In the following diagram, we can see that a region has multiple availability zones that can span multiple VPCs:

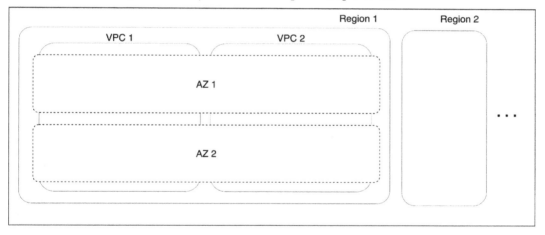

Figure 3.7 – Relationship between regions, VPCs, and availability zones

A **subnet** is a range of IP addresses associated with the VPC you have defined. A **route table** is a set of rules that determine how traffic will flow within the VPC. Every subnet you create in a VPC is automatically associated with the main route table of the VPC. A **VPC endpoint** lets you connect resources from one VPC to another and to other services.

Next, we will discuss **Classless Inter-Domain Routing** (**CIDR**) blocks and routing.

CIDR blocks and routing

CIDR is a set of standards that is useful for assigning IP addresses to a device or group of devices. A CIDR block looks like the following:

```
10.0.0.0/16
```

This defines the starting IP, and the number of IP addresses in the block. Here, the 16 means that there are 2^(32-16) or 65,536 unique addresses. When you create a CIDR block, you have to make sure that all IP addresses are contiguous, the block size is a power of 2, and IPs range from 0.0.0.0 to 256.256.256.256.

For example, the CIDR block 10.117.50.0/22 has a total of 2^(32-22), or 1,024 addresses. Now, if we would like to partition this network into four more networks with 256 addresses each, we could use the following CIDR blocks:

	10.117.50.0/24	256 addresses
	10.117.51.0/24	256 addresses
10.117.50.0.22	10.117.52.0/24	256 addresses
	10.117.53.0/24	256 addresses

Figure 3.8 – Example of using CIDR blocks to create four partitions on the network

Great, now that we know how CIDR blocks work, let us apply the same to VPCs and subnets.

Networking for HPC workloads

Referring back to *Figure 3.8*, we have made a few modifications to show CIDR blocks that define two subnets within **VPC1** in the following diagram:

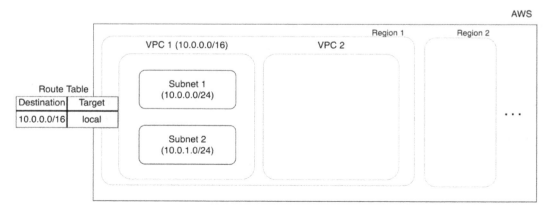

Figure 3.9 – CIDR blocks used to define two subnets within VPC1

As we can see in *Figure 3.9*, VPC 1 has a CIDR block of 10.0.0.0/16 (amounting to 65,536 addresses), and the two subnets (/24) have allocated 256 addresses each. As you have already noticed, there are several unallocated addresses in this VPC, which can be used in the future for more subnets. Routing decisions are defined using a route table, as shown in the figure. Here, each subnet is considered to be private, as traffic originating from within the VPC cannot leave the VPC. This also means that resources within this VPC cannot, by default, access the internet. One way to allow resources from within a subnet to access the internet is to add an internet gateway. For allowing only outbound internet connection from a private subnet, you can use an NAT gateway. This is often a requirement for security-sensitive workloads. This modification results in the following change to our network diagram:

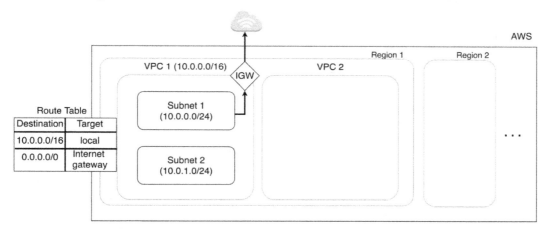

Figure 3.10 – Adding an internet gateway to Subnet 1

The main route table is associated with all subnets in the VPC, but we can also define custom route tables for each subnet. This defines whether the subnet is private, public, or VPN only. Now, if we need resources in **Subnet 2** to only access VPN resources in a corporate network via a **Virtual Private Gateway (VGW)** in *Figure 3.11*, we can create two route tables and associate them with **Subnet 1** and **Subnet 2**, as shown in the following diagram:

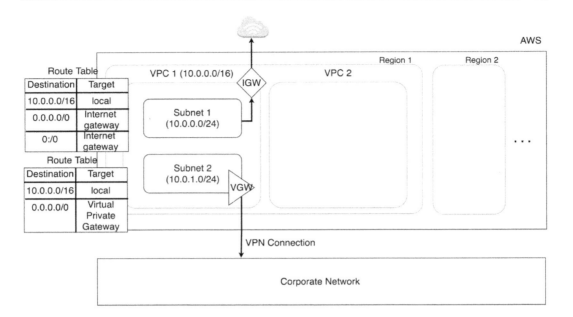

Figure 3.11 – Adding a VGW to connect to on-premises resources

A feature called **VPC peering** can be used in order to privately access resources in another VPC on AWS. With VPC peering, you can use a private networking connection between two VPCs to enable communication between them. For more information, you can visit https://docs.aws. amazon.com/vpc/latest/peering/what-is-vpc-peering.html. As shown in the following diagram, VPC peering allows resources in **VPC 1** and **VPC 2** to communicate with each other as though they are in the same network:

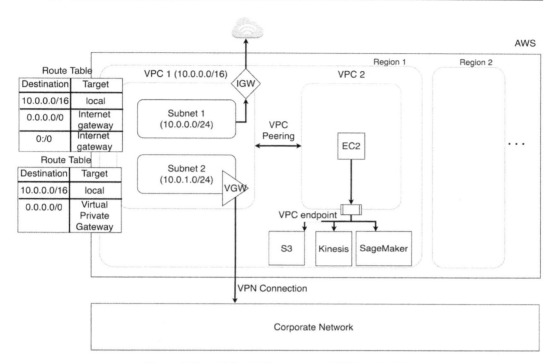

Figure 3.12 – Adding VPC peering and VPC endpoints

VPC peering can be done within VPCs in the same region or VPCs in different regions. A VPC endpoint allows resources from within a VPC (here, VPC 2) to access AWS services privately. Here, an **EC2** instance can make private API calls to services such as **Amazon S3**, **Kinesis**, or **SageMaker**. These are called **interface-type endpoints**. Gateway-type VPC endpoints are also available for Amazon S3 and DynamoDB, where you can further customize access control using policies (for example, bucket policies for Amazon S3).

Large enterprise customers with workloads that run on-premises, as well as on the cloud, may have a setup similar to *Figure 3.13*:

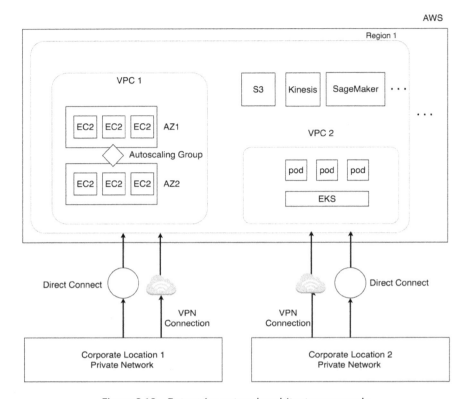

Figure 3.13 – Enterprise network architecture example

Each corporate location may be connected to AWS by using **Direct Connect** (a service for creating dedicated network connections to AWS with a VPN backup. Private subnets may host single or clusters of EC2 instances for large, permanent workloads. The cluster of EC2 instances is placed in a multi-AZ autoscaling group so that the workload can recover from the unlikely event of an AZ failure, and a minimum number of EC2 instances is maintained.

For ephemeral workloads, managed services such as EKS, Glue, or SageMaker can be used. In the preceding diagram, a private **EKS** cluster is placed in VPC 2. Since internet access is disabled by default, all container images must be local to the VPC or copied onto an ECR repository; that is, you cannot use an image from Docker Hub. To publish logs and save checkpoints, VPC endpoints are required in VPC 2 to connect to the Amazon S3 and CloudWatch services. Data stores and databases are not discussed in this diagram but are important considerations in hybrid architectures. This is because some data cannot leave the corporate network but may be anonymized and replicated on AWS temporarily.

Typically, this temporary data on AWS is used for analytics purposes before getting deleted. Lastly, hybrid architectures may also involve **AWS Outposts**, which is a fully managed service that extends AWS services, such as EC2, ECS, EKS, S3, EMR, **Relational Database Service** (**RDS**) and so on, to on-premises.

Selecting the right compute for HPC workloads

Now that you have learned about the foundations of compute and network on AWS, we are ready to explore some typical architectural patterns for compute on AWS.

Selecting the right compute for HPC and ML applications involves considering the rest of the architecture you are designing, and therefore involves all aspects of the Well-Architected Framework:

- Operational excellence
- Security
- Reliability
- Performance efficiency
- Cost optimization

We cover best practices across these pillars at the end of this section, but first, we will start with the most basic pattern of computing on AWS and add complexity as we progress.

Pattern 1 – a standalone instance

Many HPC applications that are built for simulations, financial services, CFD, or genomics can run on a single EC2 instance as long as the right instance type is selected. We discussed many of these instance-type options in the *Introducing AWS compute ecosystem* section. As shown in the following diagram, a **CloudFormation Template** can be used to launch an **EC2 Instance** in a VPC, and **Secure Shell (SSH)** access can be provided to the user for installing and using software on this instance:

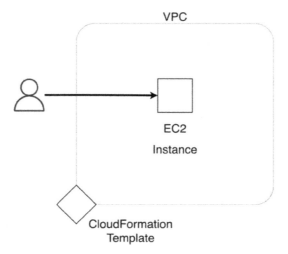

Figure 3.14 – CloudFormation Template used to launch an EC2 Instance inside a VPC

Next, we will describe a pattern that uses AWS ParallelCluster.

Pattern 2 – using AWS ParallelCluster

AWS ParallelCluster can be used to provision a cluster with head and worker nodes for massive-scale parallel processing or HPC. ParallelCluster, once launched, will be similar to on-premises HPC clusters with the added benefits of security and scalability in the cloud. These clusters can be permanent or provisioned and de-provisioned on an as-needed basis. On AWS, a user can use the AWS ParallelCluster **Command-Line Interface (CLI)** to create a cluster of EC2 instances on the fly. AWS CloudFormation is used to launch the infrastructure, including required networking, storage, and AMI configurations. As the user (or multiple users) submit jobs through the job scheduler, more instances are provisioned and de-provisioned in the autoscaling group, as shown in the following diagram:

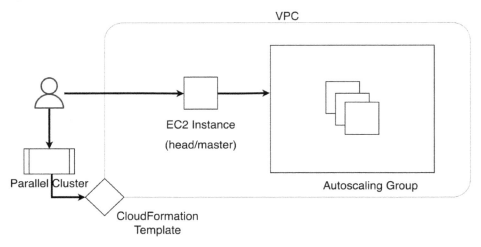

Figure 3.15 – Using AWS ParallelCluster for distributed workloads on AWS

Once the user is done with using the cluster for their HPC workloads, they can use the CLI or CloudFormation APIs to delete all resources created. As a modification to what is suggested in the following architecture, you can replace the **head/master** EC2 node with an Amazon SQS queue to get a queue-based architecture for typical HPC workloads.

Next, we will discuss how you can use AWS Batch.

Pattern 3 – using AWS Batch

AWS Batch helps run HPC and big data-based applications that are based on unconnected input configurations or files without the need to manage infrastructure. To submit a job to AWS batch, you package your application as a container and use the CLI or supported APIs to define and submit a job. With AWS Batch, you can get started quickly by using default job configurations, a built-in job queue, and integration with workflow services such as AWS Step Functions and Luigi.

As you can see in the following screenshot, the user first defines a **Docker image** (much like the image we discussed in the section on containers) and then registers this image with **Amazon ECR**. Then, the user can create a job definition in **AWS Batch** and submit one or more jobs to the job queue. Input data can be pulled from **Amazon S3**, and output data can be written to a different location on Amazon S3:

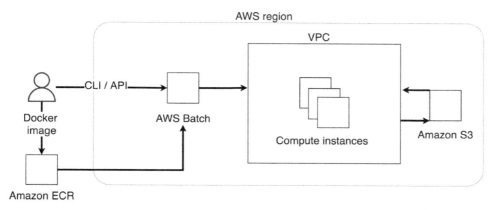

Figure 3.16 – Using AWS Batch along with AWS EC2 instances for batch workloads

Next, we will discuss patterns that help with hybrid architectures on AWS.

Pattern 4 – hybrid architecture

Customers who have already invested in large on-premises clusters, and who also want to make use of the on-demand, highly scalable, and secure AWS environment for their jobs, generally opt for a hybrid approach. In this approach, organizations decide to do one of the following:

- Run a particular job type on AWS and keep the rest on-premises
- Use AWS for overflow/excess capacity
- Use on-premises as primary data storage for security reasons, or place only the scheduler or job monitors on-premises with all of the compute being done on AWS
- Run small, test, or development jobs on-premises, but larger production jobs using high-performance or high-memory instances on AWS

On-premises data can be transferred to Amazon S3 using a software agent called DataSync (see https://docs.aws.amazon.com/datasync/latest/userguide/working-with-agents.html). Clusters that use Lustre's shared high-performance file system on-premises can make use of Amazon FSx for Lustre on AWS (for more information, see https://aws.amazon.com/fsx/lustre/). The following diagram is a reference architecture for hybrid workloads:

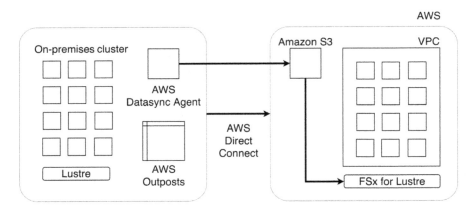

Figure 3.17 – Using FSx, S3, and AWS DataSync for hybrid architectures

Next, we will discuss patterns for container-based distributed processing

Pattern 5 – Container-based distributed processing

The following diagram is a reference architecture for container-based distributed processing workflows that are suited for HPC and other related applications:

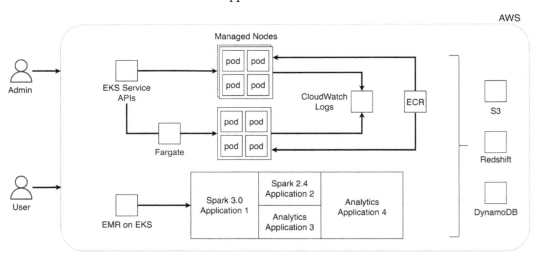

Figure 3.18 – EKS-based architecture for distributed computing

Admins can use command-line tools such as *eksctl* or CloudFormation to provision resources. Pods that are one or more containers can be run on managed EC2 nodes of your choice or via the AWS Fargate service. EMR on EKS can also be used to run open source, big data applications (for example, based on Spark) directly on EKS-managed nodes. In all of the preceding cases, containers that are provided by AWS can be used as a baseline, or completely custom containers that you build and push to ECR may be used. Applications running in EKS pods can access data from Amazon S3, Redshift, DynamoDB, or a host of other services and applications. To learn more about EKS, Fargate, or EMR on EKS, please take a look at the links provided in the *References* section.

Pattern 6 – serverless architecture

The following diagram is an example of serverless architecture that can be used for real-time, serverless processing, analytics, and business intelligence:

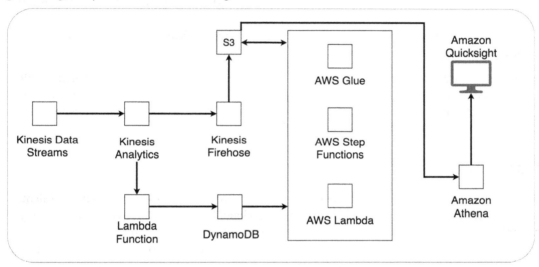

Figure 3.19 – Architecture for real-time, serverless processing and business analytics

First, Kinesis Data Streams captures data from one or more data producers. Next, **Kinesis Data Analytics** can be used to build real-time applications for transforming this incoming data using SQL, Java, Python, or Scala. Data can also be interactively processed using managed **Apache Zeppelin** notebooks (https://zeppelin.apache.org/). In this case, a **Lambda Function** is being used to continuously post-process the output of the **Kinesis Analytics** application before dropping a filtered set of results into the serverless, NoSQL database **DynamoDB**.

Simultaneously, the **Kinesis Firehose** component is being used to save incoming data into S3, which is then processed by several other serverless components such as **AWS Glu** and **AWS Lambda**, and orchestrated using **AWS Step Functions**. With AWS Glue, you can run serverless **Extract-Transform-Load** (ETL) applications that are written in familiar languages such as SQL or Spark. You can then save the output of Glue transform jobs to data stores such as Amazon S3 or Amazon Redshift. ML applications that run on Amazon SageMaker can also make use of the output data from real-time streaming analytics.

Once the data is transformed, it is ready to be queried interactively using **Amazon Athena**. Amazon Athena makes it possible for you to query data that resides in Amazon S3 using standard SQL commands. Athena is also directly integrated with the Glue Data Catalog, which makes it much easier to work with these two services without the additional burden of writing ETL jobs or scripts to enable this connection. Athena is built on the open source library **Presto** (https://prestodb.io/) and can be used to query a variety of standard formats such as CSV, JSON, Parquet, and Avro. With Athena Federated data sources, you can use a visualization tool such as **Amazon QuickSight** to run complex SQL queries.

Rather than using a dataset to visualize outputs, QuickSight, when configured correctly, can directly send these SQL queries to Athena. The results of the query can then be directly visualized interactively using multiple chart types and organized into a dashboard. These dashboards can then be shared with business analysts for further research.

In this section, we have covered various patterns around the topic of compute on AWS. Although this is not an exhaustive list of patterns, this should give you a basic idea of the components or services used and how these components are connected to each other to achieve different requirements. Next, we will describe some best practices related to HPC on AWS.

Best practices for HPC workloads

The AWS Well-Architected Framework helps with the architecting of secure, cost-effective, resilient, and high-performing applications and workloads on the cloud. It is the go-to reference when building any application. Details about the AWS Well-Architected Framework can be obtained at https://aws.amazon.com/architecture/well-architected/. However, applications in certain domains and verticals require further scrutiny and have details that need to be handled differently from the generic guidance that the AWS Well-Architected Framework provides. Thus, we have many other documents called *lenses* that provide best practice guidance; some of these lenses that are relevant to our current discussion are listed as follows:

- *Data Analytics Lens* – well-architected lens for data analytics workloads (https://docs.aws.amazon.com/wellarchitected/latest/analytics-lens/analytics-lens.html?did=wp_card&trk=wp_card)

- *Serverless Lens* – focusing on architecting serverless applications on AWS (https://docs.aws.amazon.com/wellarchitected/latest/serverless-applications-lens/welcome.html)

- *ML Lens* – for ML workloads on AWS (https://docs.aws.amazon.com/wellarchitected/latest/machine-learning-lens/welcome.html?did=wp_card&trk=wp_card)

- *HPC Lens* – focusing on HPC workloads on AWS (https://docs.aws.amazon.com/wellarchitected/latest/high-performance-computing-lens/welcome.html?did=wp_card&trk=wp_card)

While it is out of the scope of this book to go over best practices from the generic AWS Well-Architected Framework, as well as these individual lenses, we will list some common, important design considerations that are relevant to our current topic of HPC and ML:

- Both HPC and ML applications evolve over time. Organizations that freeze an architecture for several years in advance tend to be the ones that resist change and are later impacted by even larger costs to accommodate new requirements. In general, it is best practice to avoid static architectures, as the original requirements may evolve quickly. When there is a need to run more training jobs, or more HPC simulations, the architecture must allow scaling out and increase overall performance, but also return back to a steady, low-cost state when the demand is lower.

  On AWS, compute clusters can be right-sized at any given point in time, and the use of managed services can help with provisioning resources on the fly. For example, Amazon SageMaker allows users to provision various instance types for training without the undifferentiated heavy lifting of maintaining clusters or infrastructure. Customers only need to choose the framework of interest, point to training data in Amazon S3, and use the APIs to start, monitor, and stop training jobs. Customers only pay for what they use and don't pay for any idle time.

- Architecting to encourage and enable collaboration can make a significant difference to the productivity of the team running HPC and ML workloads on AWS. With teams that are becoming remote and global, the importance of effective collaboration cannot be understated. To improve collaboration, it is important to do the following:

  - Track experiments using an experiment tracking tool.

  - Enable sharing of resources such as configuration files, CloudFormation templates, pipeline definitions, code, notebooks, and data.

  - Enable automation and use tools for continuous integration, continuous delivery, continuous monitoring, continuous training, and continuous improvement.

  - Make sure that work is reproducible – this means that the inputs, environmental configuration, and packages can be easily reused and the outputs of a batch process can be verified. This helps to track changes, with audits, and to maintain high standards.

- Use ephemeral resources from managed services when possible. Again, this is applicable to both HPC and ML. When considering hybrid architectures or when migrating an on-premises workload to AWS, it is no longer necessary to completely replicate the workload on AWS. For example, running Spark-based workloads can be done on AWS Glue without the need to provision an entire EMR cluster. Similarly, you can run ML training or inference without handling the underlying clusters using Amazon SageMaker APIs.

- Consider both performance and cost when right-sizing your resources. For workloads that are not time-sensitive, using Spot instance on AWS is the simplest cost optimization strategy to follow. For HPC applications, running workloads on EC2 spot instances or using spot fleets for containerized workloads on EKS or Fargate can provide a discount of up to 90% over on-demand instances of the same type.

On SageMaker, using Spot instances is very simple – you just need to pass an argument to supported training APIs. On the other hand, for high-performance workloads, it is important to prioritize on-demand instances over spot instances so that the results of simulations or ML training jobs can be returned and analyzed in a timely manner. When choosing services or applications to use for your HPC or ML workloads, prefer pay-as-you-go pricing over licensing and upfront costs.

- Consider cost optimization and performance for the entire pipeline. It is typical for both HPC and ML applications to be designed over a pipeline – for example, data transfer, pre-processing, training or simulation, post-processing, and visualization. It is possible that some steps require less compute than others. Also, making a decision upfront about data formats or locations may force downstream steps to be more expensive in terms of time, processing resources, or cost.

- Focus on making small and frequent changes, building modular components, and testing in an automated fashion. Reducing the level of manual intervention and architecting the workload so that it does not require any downtown for maintenance is a best practice.

- For both HPC and ML, use software packages and tools that provide good documentation and support. This choice needs to be made carefully upfront, since several architectural, design, and team decisions may need to change based on this. For example, when choosing an ML framework such as PyTorch, it is important to be familiar with services on AWS that support this framework and hire a team that is well-versed in this particular framework to ensure success.

  Similarly in HPC, the choice of software that does molecular dynamics simulations will decide the scale of simulations that can be done, which services are compatible with the package on AWS, and which team members are trained and ready to make use of this software set up on AWS.

- Prioritize security and establish security best practices before beginning to develop multiple workloads or applications. These best practice areas under security are discussed in great detail in the AWS Well-Architected Framework and several of the lenses. Here, we outline the major sub-topics for completeness:

  - **Identity and Access Management (IAM)**

  - Detective controls

  - Preventive controls

  - Infrastructure protection

  - Data protection

  - Incident response

- It is normal to expect a complex architecture system to fail, but the best practice is to respond quickly and recover from these failures. Checkpointing is a common feature that is built into HPC and ML applications. A common idea is to checkpoint data or progress (or both) to a remote S3 location where a simulation or a training job can pick up after a failure. Checkpointing becomes even more important when using spot instances. When managing infrastructure on your own, you have the flexibility to deploy the application to multiple availability zones when extremely low latency requirements do not need to be met. Managed services take care of maintaining and updating instances and containers that run on these instances.

- Make sure the cluster is dynamic, can be used by multiple users simultaneously, and is designed to work over large amounts of data. In order to design the cluster successfully, use cloud native technologies to test applications and packages over a meaningful use case and not a toy problem. With the cloud, you have the ability to spin up and spin down ephemeral clusters to test out your use case at a low cost, while also making sure that a production-sized workload will work smoothly and as expected.

In this section, we have listed some best practices for HPC workloads on AWS.

Summary

In this chapter, we first described the AWS Compute ecosystem, including the various types of EC2 instances, as well as container-based services (Fargate, ECS, and EKS), and serverless compute options (AWS Lambda). We then introduced networking concepts on AWS and applied them to typical workloads using a visual walk-through. To help guide you through selecting the right compute for HPC workloads, we described several typical patterns including standalone, self-managed instances, AWS ParallelCluster, AWS Batch, hybrid architectures, container-based architectures, and completely serverless architectures for HPC. Lastly, we discussed various best practices that may further help you right-size your instances and clusters and apply the Well-Architected Framework to your workloads.

In the next chapter, we will outline the various storage services that can be used on AWS for HPC and ML workloads.

References

For additional information on the topics covered in this chapter, please navigate to the following pages:

- `https://aws.amazon.com/products/compute/`
- `https://aws.amazon.com/what-is/compute/`
- `https://aws.amazon.com/hpc/?pg=ln&sec=uc`
- `https://www.amazonaws.cn/en/ec2/instance-types/`
- `https://aws.amazon.com/ec2/instance-explorer/`
- `https://docs.aws.amazon.com/AWSEC2/latest/UserGuide/compute-optimized-instances.html`

- https://docs.aws.amazon.com/AWSEC2/latest/UserGuide/accelerated-computing-instances.html
- https://aws.amazon.com/ec2/instance-types/hpc6/
- https://aws.amazon.com/hpc/parallelcluster/
- https://aws.amazon.com/ec2/instance-types/c5/
- https://aws.amazon.com/ec2/instance-types/c6g/
- https://aws.amazon.com/ec2/instance-types/c6i/
- https://aws.amazon.com/ec2/instance-types/m5/
- https://docs.aws.amazon.com/AWSEC2/latest/UserGuide/accelerated-computing-instances.html#gpu-instances
- https://github.com/aws/aws-neuron-sdk
- https://docs.aws.amazon.com/AWSEC2/latest/UserGuide/ec2-instances-and-amis.html
- https://aws.amazon.com/ec2/instance-types/
- https://aws.amazon.com/ec2/instance-types/a1/
- https://developer.nvidia.com/blog/even-easier-introduction-cuda/
- https://aws.amazon.com/ec2/instance-types/p4/
- https://aws.amazon.com/ec2/instance-types/
- https://aws.amazon.com/ec2/instance-types/p4/
- https://www.nvidia.com/en-us/data-center/gpu-accelerated-applications/ansys-fluent/
- https://www.nvidia.com/en-in/data-center/a100/
- https://images.nvidia.com/aem-dam/en-zz/Solutions/data-center/nvidia-ampere-architecture-whitepaper.pdf
- https://pytorch.org/docs/stable/notes/cuda.html
- https://aws.amazon.com/ec2/instance-types/f1/
- https://github.com/aws/aws-fpga
- https://docs.aws.amazon.com/AWSEC2/latest/UserGuide/storage-optimized-instances.html
- https://aws.amazon.com/ec2/instance-types/i3/
- https://aws.amazon.com/ec2/instance-types/r6g/

- https://aws.amazon.com/ec2/instance-types/high-memory/
- https://aws.amazon.com/ec2/instance-types/x1/
- https://aws.amazon.com/ec2/instance-types/x1e/
- https://aws.amazon.com/ec2/instance-types/x2g/
- https://www.hpcworkshops.com/
- https://aws.amazon.com/ec2/instance-types/z1d/
- https://aws.amazon.com/containers/
- https://aws.amazon.com/ec2/?c=cn&sec=srv
- https://aws.amazon.com/containers/?nc1=f_cc
- https://aws.amazon.com/fargate/?c=cn&sec=srv
- https://aws.amazon.com/serverless/?nc2=h_ql_prod_serv
- https://aws.amazon.com/eks/?c=cn&sec=srv
- https://aws.amazon.com/ecs/?c=cn&sec=srv
- https://aws.amazon.com/lambda/
- https://aws.amazon.com/blogs/big-data/accessing-and-visualizing-data-from-multiple-data-sources-with-amazon-athena-and-amazon-quicksight/
- https://aws.amazon.com/glue/?whats-new-cards.sort-by=item.additionalFields.postDateTime&whats-new-cards.sort-order=desc
- https://docs.aws.amazon.com/kinesisanalytics/latest/dev/how-it-works-output-lambda.html
- https://aws.amazon.com/kinesis/data-analytics/
- https://aws.amazon.com/athena/?whats-new-cards.sort-by=item.additionalFields.postDateTime&whats-new-cards.sort-order=desc
- https://aws.amazon.com/emr/features/eks/
- https://docs.aws.amazon.com/emr/latest/EMR-on-EKS-DevelopmentGuide/pod-templates.html
- https://docs.aws.amazon.com/emr/latest/EMR-on-EKS-DevelopmentGuide/emr-eks.html
- https://aws.amazon.com/blogs/big-data/orchestrate-an-amazon-emr-on-amazon-eks-spark-job-with-aws-step-functions/
- https://aws.amazon.com/about-aws/global-infrastructure/regions_az/

- `https://d1.awsstatic.com/whitepapers/computational-fluid-dynamics-on-aws.pdf?cmptd_hpc3`

- `https://docs.aws.amazon.com/wellarchitected/latest/high-performance-computing-lens/general-design-principles.html`

- `https://docs.aws.amazon.com/wellarchitected/latest/machine-learning-lens/well-architected-machine-learning-design-principles.html`

- `https://docs.aws.amazon.com/outposts/latest/userguide/what-is-outposts.html`

- `https://docs.aws.amazon.com/vpc/latest/userguide/VPC_Subnets.html`

- `https://docs.aws.amazon.com/vpc/latest/userguide/vpc-nat.html`

- `https://docs.aws.amazon.com/vpc/latest/userguide/security.html`

4
Data Storage

Data storage in the cloud has become very common, not just for personal usage but also for business, computational, and application purposes as well. On the personal side, cloud storage is provided by well-known companies, ranging from a free usage tier of a few **Gigabytes (GBs)**, to pay monthly or yearly plans for **Terabytes (TBs)** of data. These services are well integrated with applications on mobile devices, enabling users to store thousands of pictures, videos, songs, and other types of files.

For applications requiring high-performance computations, cloud data storage plays an even bigger role. For example, training **Machine Learning (ML)** models over large datasets generally requires algorithms to run in a distributed fashion. If data is stored on the cloud, then it makes it much easier and more efficient for the ML platform to partition the data stored in the cloud and make these separate partitions available to the distributed components of the model training job. Similarly, for several other applications requiring large data amounts and high throughput, it makes much more sense to use cloud data storage to avoid throttling local storage. In addition, cloud data storage almost always has built-in redundancy to avoid hardware failures, accidental deletes, and hence loss of data. There are also several security and governance tools and features that are always provided with cloud data storage services. Furthermore, the true cost of ownership of data storage in the cloud is significantly reduced due to scale and infrastructure maintenance being managed by the storage service provider.

AWS provides several options for cloud data storage. In this chapter, we will learn about the various AWS data storage services, along with the security, access management, and governance aspects of these services. In addition, we will also learn about the tiered storage options to save cloud data storage costs.

We will cover the following topics in this chapter:

- AWS services for storing data
- Data security and governance
- Tiered storage for cost optimization
- Choosing the right storage option for **High-Performance Computing (HPC)** workloads

Technical requirements

The main technical requirements for being able to work with the various AWS storage options in this chapter are to have an AWS account and the appropriate permissions to use these storage services.

AWS services for storing data

AWS offers three different types of data storage services: object, file, and block. Depending on the need for the application, one or more of these types of services can be used. We will go through the AWS services spanning these storage categories in this section. The various AWS data storage services are shown in *Figure 4.1*.

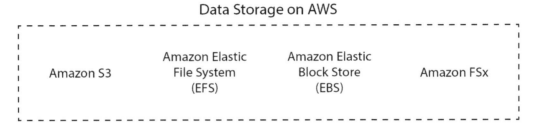

Figure 4.1 – AWS data storage services

In the next section, we will discuss the various storage options provided by AWS.

Amazon Simple Storage Service (S3)

Amazon S3 is one of the most commonly used cloud data storage services for web applications, and high-performance compute use cases. It is Amazon's object storage service providing virtually unlimited data storage. Some of the advantages of using Amazon S3 include very high scalability, durability, data availability, security, and performance. Amazon S3 can be used for a variety of cloud-native applications, ranging from simple data storage to very large data lakes to web hosting and high-performance applications, such as training very advanced and compute-intensive ML models. Amazon S3 offers several classes of storage options with differences in terms of data access, resiliency, archival needs, and cost. We can choose the storage class that best suits our use case and business needs. There is also an option for cost saving when the access pattern is unknown or changes over time (S3 Intelligent-Tiering). We will discuss these different S3 storage classes in detail in the *Tiered storage for cost optimization* section of this chapter.

Key capabilities and features of Amazon S3

In Amazon S3, data is stored as objects in *buckets*. An object is a file and any metadata that describes the file, and buckets are the resources (containers) for the objects. Some of the key capabilities of Amazon S3 are discussed next.

Data durability

Amazon S3 is designed to provide very high levels of durability to the data, up to 99.999999999%. This means that the chances of data objects stored in Amazon S3 getting lost are extremely low (average expected loss of approximately 0.000000001% of objects, or 1 out of 10,000 objects every 10 million years). For HPC applications, data durability is of the utmost importance. For example, for training an ML model, data scientists need to carry out various experiments on the same dataset in order to fine-tune the model parameters to get the best performance. If the data storage from which training and validation data is read is not durable for these experiments, then the results of the trained model will not be consistent and hence can lead to incorrect insights, as well as bad inference results. For this reason, Amazon S3 is used in many ML and other data-dependent HPC applications for storing very large amounts of data.

Object size

In Amazon S3, we can store objects up to 5 TB in size. This is especially useful for applications that require processing large files, such as videos (for example, high-definition movies or security footage), large logs, or other similar files. Many high-performance compute applications, such as training ML models for a video classification example, require processing thousands of such large files to come up with a model that makes inferences on unseen data well. A deep learning model can read these large files from Amazon S3 one (or more) at a time, store them temporarily on the model training virtual machine, compute and optimize model parameters, and then move on to the next object (file). This way, even machines with smaller disk space and memory can be used to train these computationally intensive models over large data files. Similarly, at the time of model inference, if there is a need to store the data, it can be stored in Amazon S3 for up to 5 TB of object size.

Storage classes

Amazon S3 has various storage classes. We can store data in any of these classes and can also move the data across the classes. The right storage class to pick for storing data depends on our data storage, cost, and retention needs. The different S3 storage classes are as follows:

- S3 Standard
- S3 Standard-Infrequent Access
- S3 One Zone-Infrequent Access
- S3 Intelligent-Tiering
- S3 Glacier Instant Retrieval
- S3 Glacier Flexible Retrieval
- S3 Glacier Deep Archive
- S3 Outposts

We will learn about these storage classes in the *Tiered storage for cost optimization* section of this chapter.

Storage management

Amazon S3 also has various advanced storage management options, such as data replication, prevention of accidental deletion of data, and data version control. Data in Amazon S3 can be replicated into destination buckets in the same or different AWS Regions. This can be done to add redundancy and hence reliability and also improve performance and latency. This is quite important for HPC applications as well since real-time HPC applications that need access to data stored in Amazon S3 will benefit from accessing data from a geographically closer AWS Region. Performance is generally accelerated by up to 60% when datasets are replicated across multiple AWS Regions. Amazon S3 also supports batch operations for data access, enabling various S3 operations to be carried out on billions of objects with a single API call. In addition, lifecycle policies can be configured for objects stored in Amazon S3. Using these policies, S3 objects can be moved automatically to different storage classes depending on access need, resulting in cost optimization.

Storage monitoring

Amazon S3 also has several monitoring capabilities. For example, tags can be assigned to S3 buckets, and AWS cost allocation reports can be used to view aggregated usage and cost using these tags. Amazon CloudWatch can also be used to view the health of S3 buckets. In addition, bucket- and object-level activities can also be tracked using AWS CloudTrail. *Figure 4.2* shows an example of various storage monitoring tools working with an Amazon S3 bucket:

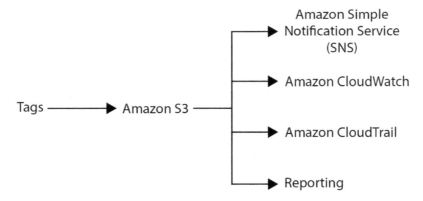

Figure 4.2 – S3 storage monitoring and management

The preceding figure shows that we can also configure Amazon **Simple Notification Service** (**SNS**) to trigger AWS Lambda to carry out various tasks in the case of certain events, such as new file uploads and so on.

Data transfer

For any application built upon large amounts of data and using S3, the data first needs to be transferred to S3. There are various services provided by AWS that work with S3 for different data transfer needs, including hybrid (premises/cloud) storage and online and offline data transfer. For example, if we want to extend our on-premise storage with cloud AWS storage, we can use **AWS Storage Gateway** (*Figure 4.3*). Some of the commonly implemented use cases for AWS Storage Gateway are the replacement of tape libraries, cloud storage backend file shares, and low-latency caching of data for on-premise applications.

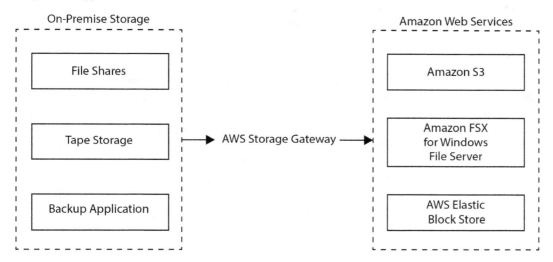

Figure 4.3 – Data transfer example using AWS Storage Gateway

For use cases requiring online data transfer, AWS DataSync can be used to efficiently transfer hundreds of terabytes into Amazon S3. In addition, AWS Transfer Family can also be used to transfer data to S3 using SFTP, FTPS, and FTP. For offline data transfer use cases, AWS Snow Family has a few options available, including AWS Snowcone, AWS Snowball, and AWS Snowmobile. For more details about the AWS Snow Family, refer to the *Further reading* section.

Performance

One big advantage of S3 for HPC applications is that it supports parallel requests. Each S3 prefix supports 3,500 requests per second to add data and 5,500 requests per second to retrieve data. Prefixes are used to organize data in S3 buckets. These are a sequence of characters at the beginning of an object's key name. We can have as many prefixes as we need in parallel, and each prefix will support this throughput. This way, we can achieve the desired throughput for our application by adding prefixes. In addition, if there is a long geographic separation between the client and the S3 bucket, we can use Amazon S3 Transfer Acceleration to transfer data. Amazon CloudFront is a globally distributed network of edge locations.

Using S3 Transfer Allocation, data is first transferred to an edge location in Amazon CloudFront. From the edge location, an optimized high-bandwidth and low-latency network path is then used to transfer the data to the S3 bucket. Furthermore, data can also be cached in CloudFront edge locations for frequently accessed requests, further optimizing performance. These performance-related features help in improving throughput and reducing latency for data access, especially suited to various HPC applications.

Consistency

Data storage requests to Amazon S3 have strong read-after-write consistency. This means that any data written (new or an overwrite) to S3 is available immediately.

Analytics

Amazon S3 also has analytics capabilities, including S3 Storage Lens and S3 Storage Class Analysis. S3 Storage Lens can be used to improve storage cost efficiency, as well as to provide best practices for data protection. In addition, it can be used to look into object storage usage and activity trends. It can provide a single view across thousands of accounts in an organization and can generate insights on various levels, such as account, bucket, and prefix. Using S3 Storage Class, we can optimize cost by deciding on when to move data to the right storage class. This information can be used to configure the lifecycle policy to make the data transfer for the S3 bucket. Amazon S3 Inventory is another S3 feature that generates daily or weekly reports, including bucket names, key names, last modification dates, object size, class, replication, encryption status, and a few additional properties.

Data security

Amazon S3 has various security measures and features. These features include blocking unauthorized users from accessing data, locking objects to prevent deletions, modifying object ownership for access control, identity and access management, discovery and protection of sensitive data, server-side and client-side encryption, the inspection of an AWS environment, and connection to S3 from on-premise or in the cloud using private IP addresses. We will learn about these data security and access management features in detail in the *Data security and governance* section.

Amazon S3 example

To be able to store data in an S3 bucket, we first need to create the bucket. Once the bucket is created, we can upload objects to the bucket. After uploading the object, we can download, move, open, or delete it. In order to create an S3 bucket, there are certain prerequisites listed as follows:

- Signing up for an AWS account
- Creating an **Identity and Access Management** (**IAM**) user or a federated user assuming an IAM role
- Signing in as an IAM user

Details of how to carry out these prerequisite steps can be found on Amazon S3's documentation web page (see the *Further reading* section).

S3 bucket creation

We can create an S3 bucket by logging into the AWS management console and selecting S3 in the services. Once in the S3 console, we will see a screen like that shown in *Figure 4.4*:

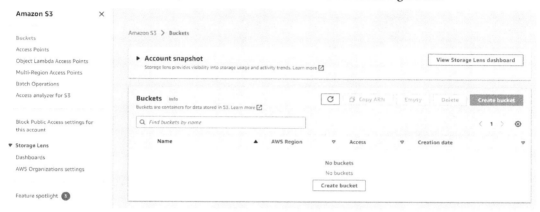

Figure 4.4 – Amazon S3 console

As *Figure 4.4* shows, we do not have any S3 buckets so far in our account. To create an S3 bucket, perform the following steps:

1. Click on one of the **Create bucket** buttons shown on this page. *Figure 4.5* shows the **Create bucket** page on the S3 console.

2. Next, we need to specify **Bucket name** and **AWS Region**. Note that the S3 bucket name needs to be globally unique:

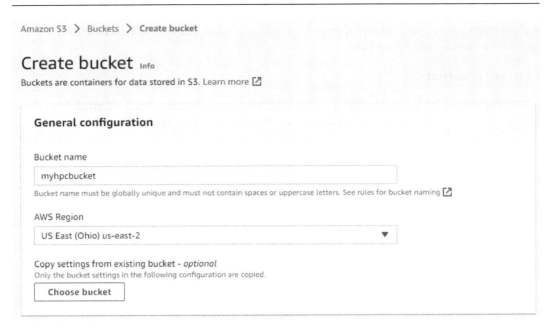

Figure 4.5 – Amazon S3 bucket creation

3.   We can also select certain bucket settings from one of our existing S3 buckets. On the S3 **Create bucket** page, we can also define whether the objects in the bucket are owned by the account creating the bucket or not, as other AWS accounts can also own the objects in the bucket. In addition, we can also select whether we want to block all public access to the bucket (as shown in *Figure 4.6*). There are also other options that we can select on the S3 bucket creation page, such as versioning, tags, encryption, and object locking:

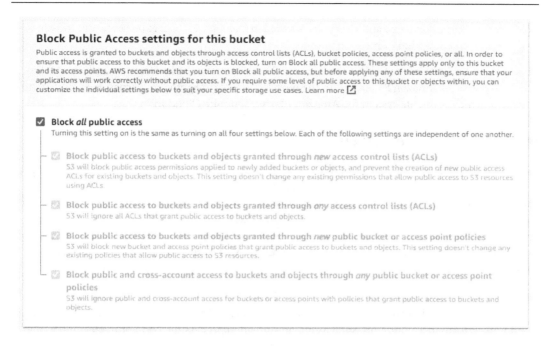

Figure 4.6 – Public access options for the S3 bucket

4. Once the bucket has been created, it will show up in the S3 console as shown in *Figure 4.7*, where we have created a bucket named `myhpcbucket`. We can add objects to it using the console, AWS **Command Line Interface** (**CLI**), AWS SDK, or Amazon S3 Rest API:

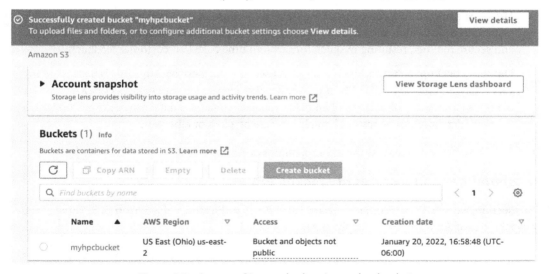

Figure 4.7 – Amazon S3 console showing myhpcbucket

We can click on the bucket name and view objects stored in it along with bucket properties, permissions, metrics, management options, and access points.

In this section, we have learned about the Amazon S3 storage class, its key features and capabilities, and how to create an Amazon S3 bucket. In the next section, we are going to discuss Amazon Elastic File System, which is the file system for Amazon Elastic Compute Cloud instances.

Amazon Elastic File System (EFS)

Amazon Elastic File System (EFS) is a fully managed serverless elastic NFS file system specifically designed for Linux workloads. It can quickly scale up to petabytes of data automatically and is well suited to work with on-premise resources as well as with various AWS services. Amazon EFS is designed such that thousands of **Amazon Elastic Compute Cloud** (**EC2**) instances can be provided with parallel shared access. In addition to EC2, EFS file systems can also be accessed by **Amazon Elastic Container Service (ECS)**, **Amazon Elastic Kubernetes Service (EKS)**, **AWS Fargate**, and **AWS Lambda** functions through a file system interface. The following are some of the common EFS use cases:

- **High-performance compute**: Since Amazon EFS is a shared file system, it is ideal for applications that require distributed workload across many instances. Applications and use cases requiring high-performance computes, such as image and video processing, content management, and ML applications, such as feature engineering, data processing, model training, numerical optimization, big data analytics, and similar applications, can benefit from Amazon EFS.

- **Containerized applications**: Amazon EFS is a very good fit for containerized applications because of its durability, which is a very important requirement of these applications. EFS integrates with Amazon container-based services such as Amazon ECS, Amazon EKS, and AWS Fargate.

- **DevOps**: Amazon EFS can be used for DevOps because of its capability to share code. This helps with code modification and the application of bug fixes and enhancements in a fast, agile, and secure manner, resulting in quick turnaround time based on customer feedback.

- **Database backup**: Amazon EFS is also often used as a database backup. This is because of the very high durability and reliability of EFS, and its low cost, along with being a POSIX-compliant file storage system – all of these often being requirements for a database backup from which the main database can be restored quickly in case of a loss or emergency.

In the next section, we will discuss the key capabilities of Amazon EFS.

Key capabilities and features of Amazon EFS

In this section, we will discuss some of the key capabilities and features of Amazon EFS. Some of the key capabilities of Amazon S3 also apply to Amazon EFS.

Durability

Like Amazon S3, Amazon EFS is also very highly durable and reliable, offering 99.999999999% durability. EFS achieves this high level of durability and redundancy by storing everything across multiple **Availability Zones** (**AZs**) within the same AWS Region (unless we select EFS One Zone storage class for the EFS storage). Because data is available across multiple AZs, EFS has the ability to recover and repair very quickly from concurrent device failures.

Storage classes

Amazon EFS also offers multiple options for storage via storage classes. These storage classes are as follows:

- Amazon EFS Standard
- Amazon EFS One Zone
- Amazon EFS Standard-Infrequent Access
- Amazon EFS One Zone-Infrequent Access

We will discuss these classes in the *Tiered storage for cost optimization* section. We can easily move files between storage classes for cost and performance optimization using policies.

Performance and throughput

Amazon EFS has two modes each for performance and throughput. For performance modes, it has **General Purpose** and **Max I/O**. General Purpose mode provides low latency for random as well as sequential input-output file system operations. Max I/O, on the other hand, is designed for very high throughput and operations per second. It is, therefore, very well suited for high-performance and highly parallelized compute applications.

For throughput, EFS has Bursting (default) and Provisioned modes. In Bursting mode, the throughput scales with the size of the file system and can burst dynamically depending on the nature of the workload. In Provisioned mode, throughput can be provisioned depending on the dedicated throughput needed by the application. It does not depend on the size of the file system.

Scalability

Amazon EFS is highly elastic and scalable. It grows up and down in size as more data is added or removed and is designed for high throughput, **Input/Output Operations Per Second** (**IOPS**), and low latency for a wide variety of workloads and use cases. It also has the capability to provide very high burst throughput for unpredictable and spiky workloads, supporting up to 10 GB/second and 500,000 IOPS at the time of writing.

AWS Backup

Amazon EFS works with AWS Backup, which is a fully managed backup service. It automates and enables us to centrally manage the EFS file systems, removing costly and tedious manual processes.

Data transfer

Like Amazon S3, Amazon EFS also works with various AWS data transfer services, such as AWS DataSync and AWS Transfer Family, for transferring data in and out of EFS for one-time migration, as well as for periodic synchronization, replication, and data recovery.

An Amazon EFS example

We can create an Amazon EFS file system either using the AWS Management Console or the AWS CLI. In this section, we will see an example of creating an EFS file system using the AWS Management Console:

1. When we log into AWS Management Console and browse to Amazon EFS service, we will see the main page like the one shown in *Figure 4.8*. It also lists all the EFS file systems that we have created in the AWS Region that we are looking at:

Figure 4.8 – Amazon EFS landing page

2. To create an Amazon EFS file system, we click on the **Create file system** button. When we click this button, we are shown a screen as shown in *Figure 4.9*. Here, we can pick any name for our file system.

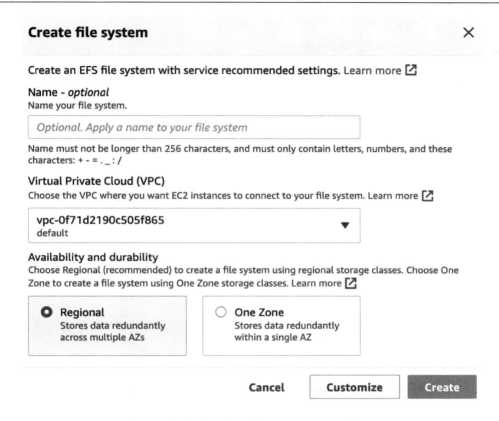

Figure 4.9 – Creating an Amazon EFS file system

3. For the file system, select a **Virtual Private Cloud** (**VPC**), and also select whether we want it across all AZs in our region or just one AZ.

4. We can also click on the **Customize** button to configure other options, such as **Lifecycle management** (*Figure 4.10*), **Performance mode** and **Throughput mode** (*Figure 4.11*), **Encryption**, **Tags**, network options, and the **File system** policy.

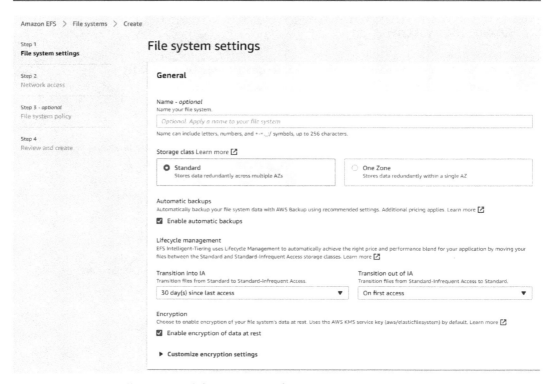

Figure 4.10 – Selecting options for an Amazon EFS file system

Figure 4.11 – Selecting additional options for Amazon EFS

5.   Clicking on **Next** and then **Create** will create the Amazon EFS file system. The EFS file system we have created is shown in *Figure 4.12*.

Figure 4.12 – Created EFS file system

6.   We can click on the file system to view its various properties as well as to get the command to mount it to a Linux instance as shown in *Figure 4.13*. Once the EFS file system is mounted, we can use it just like a regular file system.

Figure 4.13 – Mounting options and commands for the EFS file system

Let's discuss Amazon **Elastic Block Store** (**EBS**) and its key features and capabilities in the next section.

Amazon EBS

Amazon EBS is a scalable, high-performance block storage service that can be used to create storage volumes that attach to EC2 instances. These volumes can be used for various purposes, such as a regular block storage volume, creating file systems on top of these, or even running databases. The following are some of the common use cases where Amazon EBS is used:

- EBS can be used for big data analytics, where frequent resizing of clusters is needed, especially for Hadoop and Spark.

- Several types of databases can be deployed using Amazon EBS. Some examples include MySQL, Oracle, Microsoft SQL Server, Cassandra, and MongoDB.

- If we are running a computation job on an EC2 instance and need storage volume attached to it to read the data and write results without the need to scale across multiple instances, then EBS serves as a good option.

Next, let's look at some of the key features and capabilities of Amazon EBS.

Key features and capabilities of Amazon EBS

In this section, we discuss the key features and capabilities of Amazon EBS, such as volume types, snapshots, elastic volumes, EBS-optimized instances, and durability.

Volume types

EBS volumes are divided into two main categories: **SSD-backed storage** and **HDD-backed storage**. We discuss these categories here:

- **SSD-backed storage**: In SSD-backed storage, performance depends mostly on IOPS and is best suited for transactional workloads, for example, databases and boot volumes. There are two main types of SSD-backed storage volumes:

  - **Provisioned IOPS SSD volumes**: They are very high throughput volumes and are designed to provide single-digit millisecond latencies while delivering the provisioned performance 99.9% of the time. They are especially suited for critical applications that require very high uptime.

  - **General purpose SSD volumes**: These storage volumes also offer single-digit millisecond latency while delivering the provisioned performance 99% of the time. They are especially suited for transactional workloads, virtual desktops, boot volumes, and similar applications.

- **HDD-backed storage**: In HDD-backed storage, performance depends mostly on MB/s and is best suited for throughput-intensive workloads, for example, MapReduce and log processing. There are also two main types of HDD-backed storage volumes:

  - **Throughput optimized HDD volumes**: These volumes deliver performance measured in MB/s and are best suited for applications such as MapReduce, Kafka, log processing, and ETL workloads.

  - **Cold HDD volumes**: They provide the lowest cost of all EBS volumes and are backed by hard disk drives. These volumes are best suited for infrequently accessed workloads, such as cold datasets.

The option of picking the appropriate category for EBS volume provides user flexibility depending on the use case.

Snapshots

Amazon EBS has the ability to store the volume snapshots to Amazon S3. This is done incrementally, adding only the blocks of data that have been changed since the last snapshot. The data life cycle for EBS snapshots can be used to schedule the automated creation and deletion of EBS. These snapshots can be used not just for data recovery but also for initiating new volumes, expanding volume sizes, and moving EBS volumes across AZs in an AWS Region.

Elastic volumes

Using Amazon EBS Elastic Volumes, we can increase the capacity of the volume dynamically at a later point and can also change the type of volume without any downtime.

EBS-optimized instances

To fully utilize the IOPS configured for an EBS volume and provide maximum performance, some EC2 instances can be launched as EBS-optimized instances. This ensures dedicated throughput between Amazon EC2 instance and Amazon EBS volume.

Durability

Like Amazon S3 and Amazon EFS, Amazon EBS is also highly durable and reliable. Data in Amazon EBS is replicated on multiple servers in an AZ to provide redundancy and recovery in case any single component in the volume storage fails.

EBS volume creation

When creating an EC2 instance, we can add additional EBS volumes to it, as shown in *Figure 4.14* in the **Add Storage** step. In addition, we can also add new volumes from the EC2 management console after the instance has been launched. In the EC2 management console, we can review our existing volumes and snapshots along with lifecycle management policies.

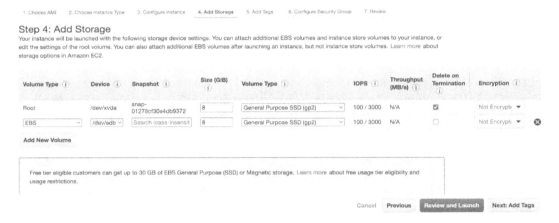

Figure 4.14 – Adding an EBS volume during the EC2 instance creation process

Figure 4.15 shows an example where we have two EBS volumes, along with their various configuration parameters:

	Name	Volume ID	Type	Size	Snapshot	Created	Availability Zone
	–	vol-0b4accb1a3866735b	gp2	8 GiB	snap-01278cf...	2022/01/26 10:52 GMT-6	us-east-2c
	–	vol-0e969b9a0bb49096f	gp2	8 GiB	–	2022/01/26 10:52 GMT-6	us-east-2c

Figure 4.15 – Volumes page in EC2 management console showing two EBS volumes that were created

So far, in this chapter, we have learned about Amazon S3, Amazon EFS, and Amazon EBS. In the next section, we discuss the Amazon FSx family of file systems.

Amazon FSx

Amazon FSx is a feature-rich, scalable, high-performance, and cost-effective family of file systems. It consists of the following commonly used four file systems:

- Amazon FSx for NetApp ONTAP
- Amazon FSx for Windows File Server
- Amazon FSx for Lustre
- Amazon FSx for OpenZFS

Some of the common examples where the FSx family of file systems is used are the following:

- FSx can deliver very low latency (sub-millisecond) and millions of IOPS and is highly scalable, making it ideal for high-performance compute applications, such as ML, numerical optimization, big data analytics, and similar applications
- Data can be migrated without breaking or modifying existing code and workflows to FSx by matching the FSx file system to that of the on-premise one
- Media and entertainment is another example where FSx is very commonly used because of it being a high-performance file system

Key features of Amazon FSx

In this section, we will discuss the key features of Amazon FSx, such as management, durability, and cost.

Fully managed

Amazon FSx is fully managed, making it very easy to migrate applications built on commonly used file systems in the industry to AWS. Linux, Windows, and macOS applications requiring very low latency and high performance work very well with Amazon FSx because of its sub-millisecond latencies.

Durability

The data in Amazon FSx is replicated across or within AZs, in addition to having the option to replicate data across AWS Regions. It also integrates with AWS Backup for backup management and protection. These features make Amazon FSx a highly available and durable family of file systems.

Cost

The cost and performance of Amazon FSx can be optimized depending on the need. It can be used for small as well as very compute-intensive workloads, such as ML and big data analytics. Like Amazon EBS, it also offers SSD and HDD storage options that can be configured for performance and storage capacity separately.

Creating an FSx file system

We can create an FSx file system by logging into the AWS management console. In the console, upon pressing the **Create file system** button, we get the option of selecting the type of FSx file system (*Figure 4.16*).

Figure 4.16 – FSx file system types

Upon selecting the type that we want to create, we are also prompted with additional options, some of which are specific to that particular FSx system being created. Some of these options, such as deployment and storage types, network and security, and Windows authentication, are shown in *Figures 4.17–4.19* for FSx for Windows File Server. These options will vary depending on the type of file system that we are creating. After creating the file system, we can mount it to our EC2 instance.

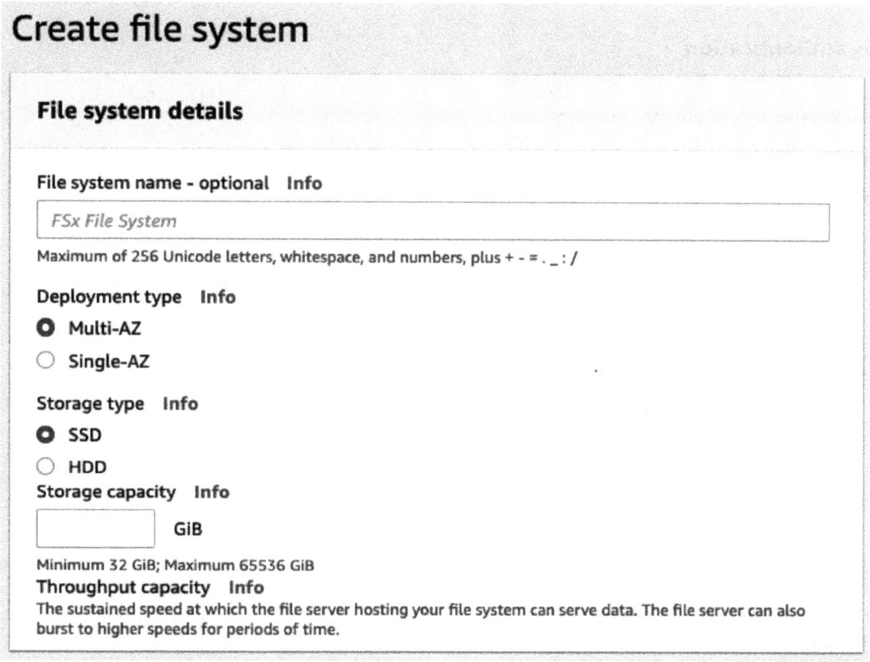

Create file system

File system details

File system name - optional   Info

FSx File System

Maximum of 256 Unicode letters, whitespace, and numbers, plus + - = . _ : /

Deployment type   Info

○ Multi-AZ

○ Single-AZ

Storage type   Info

○ SSD

○ HDD

Storage capacity   Info

GiB

Minimum 32 GiB; Maximum 65536 GiB

Throughput capacity   Info

The sustained speed at which the file server hosting your file system can serve data. The file server can also burst to higher speeds for periods of time.

Figure 4.17 – Creating FSx for Windows File Server

Network & security

Virtual Private Cloud (VPC)   Info
Specify the VPC from which your file system is accessible.

Default VPC | vpc-0f71d2190c505f865   ▼

VPC Security Groups   Info
Specify VPC Security Groups to associate with your file system's network interfaces.

Choose VPC security group(s)   ▼

sg-09515e41c6fa5df6c (default) ✕

Preferred subnet   Info
Specify the preferred subnet for your file system.

subnet-03f2e928dc3336222 (us-east-2c)   ▼

Standby subnet

subnet-0bd697fd428792c67 (us-east-2b)   ▼

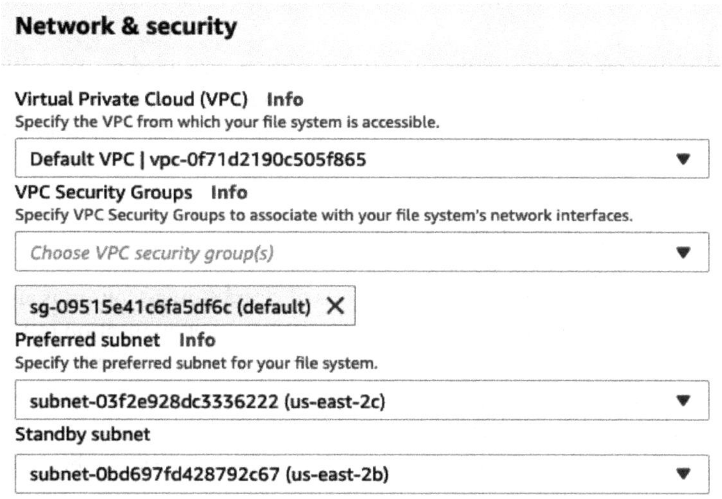

Figure 4.18 – Network and security options for FSx for Windows File Server

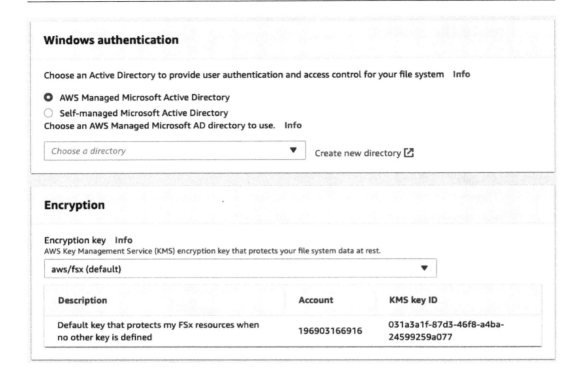

Figure 4.19 – Authentication and encryption options for FSx for Windows File Server

This concludes our discussion on the various data storage options provided by AWS. We have learned about Amazon S3, Amazon EFS, Amazon EBS, and the Amazon FSx family, along with their key capabilities and features. In the next section, we will learn about the data security and governance aspects of cloud data storage on AWS.

Data security and governance

Data security and governance are very important aspects of cloud storage solutions and applications, whether they are web pages, file storage, ML applications, or any other application utilizing cloud data storage. It is of absolute importance that the data is protected while being at rest in storage or in transit. In addition, access controls should be applied to different users based on the privileges needed for data access. All the AWS data storage services mentioned previously have various security, protection, access management, and governance features. We will discuss these features in the following sections.

IAM

In order to be able to access AWS resources, we need an AWS account and to authenticate every time we log in. Once we have logged into AWS, we need permission to access the AWS resources. AWS IAM is used to attach permission policies to users, groups, and roles. These permission policies govern and control the access to AWS resources. We can specify who can access which resource and what actions they can take for that resource (for example, creating an S3 bucket, adding objects, listing objects, deleting objects, and so on). All AWS data storage services described in the previous section are integrated with AWS IAM, and access to these services and associated resources can be governed and controlled using IAM.

Data protection

All AWS data storage and file system services have various data protection features. We should always protect AWS account credentials and create individual user accounts using AWS IAM, giving each user the least privileges to access AWS resources that are needed for their job duties. A few additional security recommendations that help with data protection in AWS are the use of **Multi-Factor Authentication (MFA)**, **Secure Socket Layer (SSL)**, **Transport Layer Security (TLS)**, activity logging via **AWS CloudTrail**, AWS encryption solutions, and **Amazon Macie** for Amazon S3 for securing personal and sensitive data.

The various tiers of Amazon S3 (except One Zone-IA) store all data objects across at least three AZs in an AWS Region. The One Zone-IA storage class provides redundancy and protection by storing data on multiple devices within the same AZ. In addition, with the help of versioning, different versions of data objects can be preserved and recovered as needed.

Data encryption

AWS provides data encryption at rest as well as in transit. Encryption in transit can be carried out by enabling SSL/TLS. Also, all data across AWS Regions flowing over the AWS global network is automatically encrypted. In Amazon S3, data can also be encrypted using client-side encryption. Data at rest on Amazon S3 can be encrypted using either server-side encryption or client-side encryption. For server-side encryption, we can either use **AWS Key Management Service (KMS)** or Amazon S3-managed encryption. For client-side encryption, we need to take care of the encryption process and upload objects to S3 after encrypting them. We can also create encrypted Amazon EFS file systems, Amazon EBS volumes, and the Amazon FSx family of file systems.

Logging and monitoring

Logging and monitoring are two very essential components of any storage solution since we need to keep track of who is accessing the data and what they are doing with it. Often this logging is also necessary to satisfy audits and build analytics and reports for usage and threat analysis. AWS data storage services have several logging and monitoring tools available. Amazon CloudWatch alarms can be used to monitor metrics, triggering alarms to Amazon SNS to send notifications. We can also use Amazon CloudTrail to view actions taken by a user, IAM roles, and AWS services in our data storage and file systems. In addition, there are other logging and monitoring features, such as Amazon CloudWatch Logs to access log files and Amazon CloudWatch Events to capture state information and take corrective actions.

Resilience

AWS infrastructure consists of AWS Regions, which in turn consist of multiple physically separated AZs. These AZs have high throughput and low latency connectivity. By default, S3, EFS, EBS, and FSx resources are backed up and replicated either across multiple AZs across an AWS Region or within the same AZ, depending on the configuration picked by the user. In addition, there are also several other resilience features specific to these data storage and file systems, such as lifecycle configuration, versioning, S3 object lock, EBS snapshots, and so on. All these features make the data stored on AWS highly resilient to failures and loss.

In addition to the aforementioned mentioned security and governance features of data storage services on AWS, there are several other options available as well, such as the configuration of VPCs, network isolation, and a few additional options. With the combination of these tools and resources, AWS data storage services are among the most secure cloud storage services available.

AWS provides the option of using tiered storage for Amazon S3 and Amazon EFS. This helps with optimizing the data storage cost for the user.

Tiered storage for cost optimization

AWS provides options for configuring its data storage services with various different tiers of storage types. This significantly helps with optimizing cost and performance depending on the use case requirements. In this section, we will discuss the tiered storage options for Amazon S3 and Amazon EFS.

Amazon S3 storage classes

As mentioned in the *Amazon Simple Storage Service (S3)* section, there are various storage classes depending on the use case, access pattern, and cost requirements. We can configure S3 storage classes at the object level. We will discuss these storage classes in the following sections.

Amazon S3 Standard

Amazon S3 Standard is the general-purpose S3 object storage commonly used for frequently accessed data. It provides high throughput and low latency. Some of the common applications of S3 Standard are online gaming, big data analytics, ML model training and data storage, an offline feature store for ML applications, content storage, and distribution, and websites with dynamic content.

Amazon S3 Intelligent-Tiering

Amazon S3 Intelligent-Tiering is the storage class for unknown, unpredictable, and changing access patterns. There are three access tiers in S3 Intelligent-Tiering – frequent, infrequent, and archive tiers. S3 Intelligent-Tiering monitors access patterns and moves data to the appropriate tiers accordingly in order to save costs without impacting performance, retrieval fees, or creating operational overhead. In addition, we can also set up S3 Intelligent-Tiering to move data to the Deep Archive Access tier for data that is accessed very rarely (180 days or more). This can result in further additional cost savings.

Amazon S3 Standard-Infrequent Access

Amazon S3 Standard-Infrequent Access is for use cases where data is generally accessed less frequently, but rapid access may be required. It offers a low per GB storage price and retrieval charge but the same performance and durability as S3 Standard. Some of the common use cases for this tier are backups, a data store for disaster recovery, and long-term storage. For high-performance compute applications, such as ML, this storage tier can be used to store historical data on which models have already been trained or analytics have already been carried out and is not needed for model retraining for a while.

Amazon S3 One Zone-Infrequent Access

Amazon S3 One Zone-Infrequent Access is very similar to Amazon S3 Standard-Infrequent Access, but the data is stored in only one AZ (multiple devices) instead of the default three AZs within the same AWS Region as for other S3 storage classes. This is even more cost-effective than the S3 Standard-Infrequent Access storage class and is commonly used for storing secondary backups or easily re-creatable data, for example, engineered features no longer used for active ML model training.

Amazon S3 Glacier

Amazon S3 Glacier storage classes are highly flexible, low-cost, and high-performance data archival storage classes. In Amazon S3 Glacier, there are three storage classes. Amazon S3 Glacier Instant Retrieval is generally used where data is accessed very rarely, but the retrieval is required with latency in milliseconds, for example, news media assets and genomics data. Amazon S3 Flexible Retrieval is for use cases where large datasets such as backup recovery data need to be retrieved at no additional cost, but instant retrieval is not a requirement. The usual retrieval times for such use cases are a few minutes to a few hours. Amazon S3 Glacier Deep Archive is for use cases that require very infrequent retrieval, such as preserved digital media and compliance archives, for example. It is the lowest-cost storage of all the options discussed previously, and the typical retrieval time is 12 hours to 2 days.

S3 on Outposts

For on-premise AWS Outposts environments, object storage can be configured using **Amazon S3 on Outposts**. It stores data reliably and redundantly across multiple devices and servers on AWS Outposts, especially suited for use cases with local data residency requirements.

In the following section, we will discuss the different storage classes for Amazon EFS.

Amazon EFS storage classes

Amazon EFS provides the option of different storage classes for access based on how frequently the data needs to be accessed, as discussed here.

Amazon EFS Standard and EFS Standard-Infrequent Access classes

Amazon EFS Standard and EFS Standard-Infrequent Access classes are highly available, durable, and elastic file system storage classes. They are both replicated across multiple geographically separated AZs within an AWS Region. EFS Standard is for use cases where our data needs frequent access, whereas the EFS Standard-Infrequent Access class is for use cases where frequent data access is not required. Using EFS Standard-Infrequent Access, we can reduce the storage cost significantly.

Amazon EFS One Zone and EFS One Zone-Infrequent Access classes

Amazon EFS One Zone and EFS One Zone-Infrequent Access classes store data within a single AZ across multiple devices, reducing the storage cost compared to Amazon EFS Standard and EFS Standard-Infrequent Access classes, respectively. For frequently accessed files, EFS One Zone is recommended, whereas for infrequently accessed files, EFS One Zone-Infrequent Access class is recommended.

With multiple options available for Amazon S3 and Amazon EFS, the right approach is to first determine performance, access, and retrieval needs for a use case and then pick the storage class that satisfies these requirements while minimizing the total cost. Significant amounts of savings can be achieved if we pick the right storage class, especially for very big datasets and use cases where data scales significantly over time.

So far, we have discussed various AWS storage options for HPC along with their capabilities and cost optimization options. In the next section, we will learn about how to pick the right storage option for our HPC use cases.

Choosing the right storage option for HPC workloads

With so many choices available for cloud data storage, it becomes challenging to decide which storage option to pick for HPC workloads. The choice of data storage depends heavily on the use case and performance, throughput, latency, scaling, archival, and retrieval requirements.

For use cases where we need to archive our object data for a very long time, Amazon S3 should be considered. In addition, Amazon S3 can be very well suited to several HPC applications since it can be accessed by other AWS services. For example, in Amazon SageMaker, we can carry out feature engineering using data stored in Amazon S3 and then ingest those features in the SageMaker offline feature store, which is, again, stored in Amazon S3. Amazon SageMaker uses Amazon S3 for ML model training. It reads data from Amazon S3 and carries out model fitting, hyperparameter optimization, and validation using this data. The model artifacts created as a result are then stored in Amazon S3 as well, which can be used for real-time or batch inference. In addition to ML, Amazon S3 is also a good choice of storage for carrying out data analytics, for storing data on which we want to run complex queries, data archiving, and backups.

Amazon EFS is a shared file system for Amazon EC2 instances. Thousands of EC2 instances can share the same EFS file system. This makes it ideal for applications where high-performance scaling is needed. High-performance compute applications such as content management systems, distributed applications running on various instances needing access to the same data in the file system, and very large-scale data analysis are a few examples where EFS should be used.

Amazon EBS is a block storage service for single EC2 instances (except EBS Multi-Attach), so the main use case for EBS is when we need high-performance storage for an EC2 instance. For high-performance compute applications such as ML and numerical optimization, often we need to access data for training and tuning our algorithms. The size of the data is often very large to be stored in memory during the process (for example, several thousand videos). In such cases, we may store data temporarily on EBS drives attached to the EC2 instances on which we are running our algorithms, making it much faster to swap and read between data files to carry out the compute operations.

Amazon FSx should be used when we have a similar file system running on-premise and we want to migrate our applications to the cloud while also designing new applications on a similar file system without worrying about underlying infrastructure and tools and process changes.

These are some of the examples we have discussed here where high-performance compute applications can benefit from various AWS data storage options. At the time of designing the architecture, it is very important that we pick the right selection of storage options to make our application give the best performance while also making sure that we do not incur unneeded costs.

Summary

In this chapter, we have discussed the different data storage options available on AWS, along with their main features and capabilities. We have introduced Amazon S3 – a highly scalable and reliable object storage service, Amazon EFS – a shared file system for EC2 instances, Amazon EBS – block storage for EC2 instances, and the Amazon FSx family of file systems. We have also talked about the data protection and governance capabilities of these services and how they integrate with various other data protection, access management, encryption, logging, and monitoring services. We have also explored the various tiers of storage available for Amazon S3 and Amazon EFS and how we can use these tiers to optimize cost for our use cases. Finally, we have discussed a few examples of when to use which data storage service for high-performance compute applications.

Now that we have a good understanding of various AWS data storage services, we are ready to move on to the next part of the book, *Chapter 5, Data Analysis*, which begins with how to carry out data analysis using AWS services.

Further reading

For further reading on the material we learned in this chapter, please refer to the following articles:

- *Amazon S3 Features*: `https://aws.amazon.com/s3/features/`
- *Amazon Regions and Availability Zones*: `https://aws.amazon.com/about-aws/global-infrastructure/regions_az/`
- *AWS Snow Family*: `https://aws.amazon.com/snow/`
- *Getting started with Amazon S3*: `https://docs.aws.amazon.com/AmazonS3/latest/userguide/GetStartedWithS3.html`
- *Prerequisite: Setting up Amazon S3*: `https://docs.aws.amazon.com/AmazonS3/latest/userguide/setting-up-s3.html`
- *AWS EFS Deep Dive: What is it and when to use it*: `https://www.learnaws.org/2021/01/23/aws-efs-deep-dive/`
- *Achieve highly available and durable database backup workflows with Amazon EFS*: `https://aws.amazon.com/blogs/storage/using-amazon-efs-to-cost-optimize-highly-available-durable-database-backup-workflows/`
- *Amazon EFS features*: `https://aws.amazon.com/efs/features/`
- *AWS Backup*: `https://aws.amazon.com/backup/`
- *Amazon EBS features*: `https://aws.amazon.com/ebs/features/`
- *Amazon FSx Documentation*: `https://docs.aws.amazon.com/fsx/index.html`
- *Amazon S3 Glacier storage classes*: `https://aws.amazon.com/s3/storage-classes/glacier/`

Part 2: Applied Modeling

Part 2 focuses on the application of **high-performance computing** (**HPC**) for machine learning. It includes a hands-on implementation of an end-to-end solution starting with analyzing large amounts of data and then covering distributed training and deploying models at scale, including performance optimization and machine learning at the edge.

This part comprises the following chapters:

- *Chapter 5, Data Analysis*
- *Chapter 6, Distributed Training of Machine Learning Models*
- *Chapter 7, Deploying Machine Learning Models at Scale*
- *Chapter 8, Optimizing and Managing Machine Learning Models for Edge Deployment*
- *Chapter 9, Performance Optimization for Real-Time Inference*
- *Chapter 10, Data Visualization*

5

Data Analysis

One of the fundamental principles behind any large-scale data science procedure is the simple fact that any **Machine Learning (ML)** model produced is only as good as the data on which it is trained. Beginner data scientists often make the mistake of assuming that they just need to find the right ML model for their use case and then simply train or fit the data to the model. However, nothing could be further from the truth. Getting the best possible model requires exploring the data, with the goal being to fully understand the data. Once the data scientist understands the data and how the ML model can be trained on it, the data scientist often spends most of their time further cleaning and modifying the data, also referred to as wrangling the data, to prepare it for model training and building.

While this data analysis task may seem conceptually straightforward, the task becomes far more complicated when we factor in the *type* (images, text, tabular, and so on) and the *amount/volume* of data we are exploring. Furthermore, where the data is stored and getting access to it can also make the exercise even more overwhelming for the data scientist. For example, useful ML data may be stored within a data warehouse or located within various relational databases, often requiring various tools or programmatic API calls to mine the right data. Likewise, key information may be located across multiple file servers or within various buckets of a cloud-based object store. Locating the data and ensuring the correct permissions to access the data can further delay a data scientist from getting started.

So, with these challenges in mind, in this chapter, we will review some practical ways to explore, understand, and essentially wrangle different types, as well as large quantities of data, to train ML models. Additionally, we will examine some of the capabilities and services that AWS provides to make this task less daunting.

Therefore, this chapter will cover the following topics:

- Exploring data analysis methods
- Reviewing AWS services for data analysis
- Analyzing large amounts of structured and unstructured data
- Processing data at scale on AWS

Technical requirements

You should have the following prerequisites before getting started with this chapter:

- Familiarity with AWS services and their basic usage.

- A web browser (for the best experience, it is recommended that you use a Chrome or Firefox browser).

- An AWS account (if you are unfamiliar with how to get started with an AWS account, you can go to this link: `https://aws.amazon.com/getting-started/`).

- Familiarity with the AWS Free Tier (the Free Tier will allow you to access some of the AWS services for free, depending on resource limits. You can familiarize yourself with these limits at this link: `https://aws.amazon.com/free/`).

- Example Jupyter notebooks for this chapter are provided in the companion GitHub repository (`https://github.com/PacktPublishing/Applied-Machine-Learning-and-High-Performance-Computing-on-AWS/tree/main/Chapter05`).

Exploring data analysis methods

As highlighted at the outset of this chapter, the task of gathering and exploring these various sources of data can seem somewhat daunting. So, you may be wondering at this point *where and how to begin the data analysis process?* To answer this question, let's explore some of the methods we can use to analyze your data and prepare it for the ML task.

Gathering the data

One of the first steps to getting started with a data analysis task is to gather the relevant data from various silos into a specific location. This single location is commonly referred to as a data lake. Once the relevant data has been co-located into a single data lake, the activity of moving data in or out of the lake becomes significantly easier.

For example, let's imagine for a moment that a data scientist is tasked with building a product recommendation model. Using the data lake as a central store, they can query a customer database to get all the customer-specific data, typically from a relational database or a data warehouse, and then combine the customer's clickstream data from the web application's transaction logs to get a common source of all the required information to predict product recommendations. Moreover, by sourcing product-specific image data from the product catalog, the data scientist can further explore the various characteristics of product images that may enhance or contribute to the ML model's predictive potential.

So, once that holistic dataset is gathered together and stored in a common repository or data store, we can move on to the next approach to data analysis, which is understanding the data structure.

Understanding the data structure

Once the data has been gathered into a common location, before a data scientist can fully investigate how it can be used to suggest an ML hypothesis, we need to understand the structure of the data. Since the dataset may be created from multiple sources, understanding the structure of the data is important before it can be analyzed effectively.

For instance, if we continue with the product recommendation example, the data scientist may work with structured customer data in the form of a tabular dataset from a relational database or data warehouse. Added to this, when pulling the customer interaction data from the web servers, the data scientist may work with time series or JSON formatted data, commonly referred to as semi-structured data. Lastly, when incorporating product images into the mix, the data scientist will deal with image data, which is an example of unstructured data.

So, understanding the nature or structure of the data determines how to extract the critical information we need and analyze it effectively. Moreover, knowing the type of data we're dealing with will also influence the type of tools, such as **Application Programming Interfaces** (**APIs**), and even the infrastructure resources required to understand data.

> **Note**
> We will be exploring these tools and infrastructure resources in depth further on in the chapter.

Once we understand the data structure, we can apply this understanding to another technique of data analysis, that is, describing the data itself.

Describing the data

Once we understand the data's structure, we can describe or summarize the characteristics of the data to further explore how these characteristics influence our overall hypothesis. This methodology is commonly referred to as applying **descriptive statistics** to the data, whereby a data scientist will try to describe and understand the various features of the dataset by summarizing the collective properties of each feature within the data in terms of centrality, variability, and data counts.

Let's explore what each of these terms means to see how they can be used to describe the data.

Determining central tendency

By using descriptive statistics to summarize the central tendency of the data, we are essentially focusing on the average, middle, or center position within the distribution of a specific feature of the dataset. This gives the data scientist an idea of what is normal or average about a feature of the dataset, allowing them to compare these averages with other features of the data or even the entirety of the data.

For example, let's say that customer A visited our website 10 times a day but only purchased 1 item. By comparing the average visits of customer A with the total number of customer visits, we can see how customer A ranks in comparison. If we then compare the number of items purchased with the total number of items, we can further gauge what is considered normal based on the customer's ranking.

Measuring variability

Using descriptive statistics to measure the variability of the data or how the data is spread is extremely important in ML. Understanding how the data is distributed will give the data scientist a good idea of whether the data is proportional or not. For example, in the product recommendation use case where we have data with a greater spread of customers who purchase books versus customers who purchase lawnmowers. In this case, when a model is trained on this data, it will be biased toward recommending books over lawnmowers.

Counting the data points

Not to be confused with dataset sizes, dataset counts refer to the number or quantity of individual data points within the dataset. Summarizing the number of data points for each feature within the dataset can further help the data scientist to verify whether they have an adequate number of data points or observations for each feature. Having sufficient quantities of data points will further help to justify the overall hypothesis.

Additionally, by comparing the individual quantities of data points for each feature, the data scientist can determine whether there are any missing data points. Since the majority of ML algorithms don't deal well with missing data, the data scientist can circumvent any unnecessary issues during the model training process by dealing with these missing values during the data analysis process.

> **Note**
> While these previously shown descriptive techniques can help us understand the characteristics of the data, a separate branch of statistics, called **inferential statistics**, is also required to measure and understand how features interact with one another within the entire dataset. This factor is important when dealing with large quantities of data where inferential techniques will need to be applied if we don't have a mechanism to analyze large datasets at scale.

We've all heard the saying that *a picture paints a thousand words*. So, once we have a good understanding of the dataset's characteristics, another important data analysis technique is to visualize these characteristics.

Visualizing the data

While summarizing the characteristics of the data provides useful information to the data scientist, we are essentially adding more data to the analysis task. Plotting or charting this additional information can potentially reveal further characteristics of the data that summary and inferential statistics may miss.

For example, using visualization to understand the variance and spread of the data points, a data scientist may use a bar chart to group the various data points into *bins* with equal ranges to visualize the distribution of data points in each *bin*. Furthermore, by using a boxplot, a data scientist can visualize whether there are any outlying data points that influence the overall distribution of the data.

Depending on the type of data and structure of the dataset, many different types of plots and charts can be used to visualize the data. It is outside the scope of this chapter to dive into each and every type of plot available and how it can be used. However, it is sufficient to say that data visualization is an essential methodology for exploratory data analysis to verify data quality and help the data scientist become more familiar with the structure and characteristics of the dataset.

Reviewing the data analytics life cycle

While there are many other data analysis methodologies, of which we have only touched on four, we can summarize the overall data analysis methodology in the following steps:

1. Identify the use case and questions that need to be answered from the data, plus the features the ML model needs to predict.

2. Gather or mine the data into a common data store.

3. Explore and describe the data.

4. Visualize the data.

5. Clean the data and prepare it for model training, plus account for any missing data.

6. Engineer new features to enhance the hypothesis and improve the ML model's predictive capability.

7. Rinse and repeat to ensure that the data, as well as the ML model, addresses the business use case.

Now that we have reviewed some of the important data analysis methodologies and the analysis life cycle, let's review some of the capabilities and services that AWS provides to apply these techniques, especially at scale.

Reviewing the AWS services for data analysis

AWS provides multiple services that are geared to help the data scientist analyze either structured, semi-structured, or unstructured data at scale. A common style across all these services is to provide users with the flexibility of choice to match the right aspects of each service as it applies to the use case. At times, it may seem confusing to the user which service to leverage for their use case.

Thus, in this section, we will map some of the AWS capabilities to the methodologies we've reviewed in the previous section.

Unifying the data into a common store

To address the requirement of storing the relevant global population of data from multiple sources in a common store, AWS provides the Amazon **Simple Storage Service** (**S3**) object storage service, allowing users to store structured, semi-structured, and unstructured data as objects within buckets.

> **Note**
>
> If you are unfamiliar with the S3 service, how it works, and how to use it, you can review the S3 product page here: `https://aws.amazon.com/s3/`.

Consequently, S3 is the best place to create a data lake as it has unrivaled security, availability, and scalability. Incidentally, S3 also provides multiple additional resources to bring data into the store.

However, setting up and managing data lakes can be time-consuming and intricate, and may take up to several weeks to set it up, based on your requirements. It often requires loading data from multiple different sources, setting up partitions, enabling encryption, and providing auditable access. Subsequently, AWS provides **AWS Lake Formation** (`https://aws.amazon.com/lake-formation`) to build secure data lakes in mere days.

Creating a data structure for analysis

As was highlighted in the previous section, understanding the underlying structure of our data is critical to extracting the key information needed for analysis. So, once our data is stored in S3, we can leverage **Amazon Athena** (`https://aws.amazon.com/athena`) using **Structured Query Language** (**SQL**), or leverage **Amazon EMR** (`https://aws.amazon.com/emr/`) to analyze large-scale data, using open source tooling, such as **Apache Spark** (`https://spark.apache.org/`) and the **PySpark** (`https://spark.apache.org/docs/latest/api/python/index.html?highlight=pyspark`) Python interface. Let's explore these analytics services further by starting with an overview of Amazon Athena.

Reviewing Amazon Athena

Athena makes it easy to define a schema, a conceptual design of the data structure, and query the structured or semi-structured data in S3 using SQL, making it easy for the data scientist to obtain key information for analysis on large datasets.

One critical aspect of Athena is the fact that it is **serverless**. This is of key importance to data scientists because there are no requirements for building and managing infrastructure resources. This means that data scientists immediately start their analysis tasks without needing to rely on a platform or infrastructure team to develop and build an analytics architecture.

However, the expertise to perform SQL queries may or may not be within a data scientist's wheelhouse since the majority of practitioners are more familiar with Python data analysis tools, such as **pandas** (`https://pandas.pydata.org/`). This is where Spark and EMR come in. Let's review how Amazon EMR can help.

Reviewing Amazon EMR

Amazon **EMR** or **Elastic MapReduce** is essentially a managed infrastructure provided by AWS, on which you can run Apache Spark. Since it's a managed service, EMR allows the infrastructure team to easily provision, manage, and automatically scale large Spark clusters, allowing data scientists to run petabyte-scale analytics on their data using tools they are familiar with.

There are two key points to be aware of when leveraging EMR and Spark for data analysis. Firstly, unlike Athena, EMR is not serverless and requires an infrastructure team to provision and manage a cluster of EMR nodes. While these tasks have been automated when using EMR, taking between 15 to 20 minutes to provision a cluster, the fact still remains that these infrastructure resources require a build-out before the data scientist can leverage them.

Secondly, EMR with Spark makes use of **Resilient Distributed Datasets (RDDs)** (`https://spark.apache.org/docs/3.2.1/rdd-programming-guide.html#resilient-distributed-datasets-rdds`) to perform petabyte-scale data analysis by alleviating the memory limitations often imposed when using pandas. Essentially, this allows the data scientist to perform analysis tasks on the entire population of data as opposed to extracting a small enough sample to fit into memory, performing the descriptive analysis tasks on the said sample, and then inferring the analysis back onto the global population. Having the ability to execute an analysis of all of the data in a single step can significantly reduce the time taken for the data scientist to describe and understand the data.

Visualizing the data at scale

As if ingesting and analyzing large-scale datasets isn't complicated enough for a data scientist, using programmatic visualization libraries such as Matplotlib (`https://matplotlib.org/`) and Seaborn (`https://seaborn.pydata.org/`) can further complicate the analysis task.

So, in order to assist data scientists in visualizing data and gaining additional insights plus performing both descriptive and inferential statistics, AWS provides the **Amazon QuickSight** (`https://aws.amazon.com/quicksight/`) service. QuickSight allows data scientists to connect to their data on S3, as well as other data sources, to create interactive charts and plots.

Furthermore, leveraging QuickSight for data visualization tasks does not require the data scientist to rely on their infrastructure teams to provision resources as QuickSight is also serverless.

Choosing the right AWS service

As you can imagine, AWS provides many more services and capabilities for large-scale data analysis, with S3, Athena, and QuickSight being only a few of the more common technologies that specifically focus on data analytics tasks. Choosing the right capability is dependent on the use case and may require integrating other infrastructure resources. The key takeaway from this brief introduction to these services is that, where possible, data scientists should not be burned by having to manage resources outside of the already complicated task of data analysis.

Therefore, from the perspective of the data scientist or ML practitioner, AWS provides a dedicated service with capabilities specifically dedicated to common ML tasks called **Amazon SageMaker** (`https://aws.amazon.com/sagemaker/`).

Therefore, in the next section, we will demonstrate how SageMaker can help with analyzing large-scale data for ML without the data scientist having to personally manage or rely on infrastructure teams to manage resources.

Analyzing large amounts of structured and unstructured data

Up until this point in the chapter, we have reviewed some of the typical methods for large-scale data analysis and introduced some of the key AWS services that focus on making the analysis task easier for users. In this section, we will practically introduce Amazon SageMaker as a comprehensive service that allows both the novice as well as the experienced ML practitioner to perform these data analysis tasks.

While SageMaker is a fully managed infrastructure provided by AWS along with tools and workflows that cater to each step of the ML process, it also offers a fully **Integrated Development Environment** (**IDE**) specifically for ML development called **Amazon SageMaker Studio** (`https://aws.amazon.com/sagemaker/studio/`). SageMaker Studio provides a data scientist with the capabilities to develop, manage, and view each part of the ML life cycle, including exploratory data analysis.

But, before jumping into a hands-on example where we can perform large-scale data analysis using Studio, we need to configure a SageMaker domain. A SageMaker Studio domain comprises a set of authorized data scientists, pre-built data science tools, and security guard rails. Within the domain, these users can share access to AWS analysis services, ML experiment data, visualizations, and Jupyter notebooks.

Let's get started.

Setting up EMR and SageMaker Studio

We will use an **AWS CloudFormation** (`https://aws.amazon.com/cloudformation/`) template to perform the following tasks:

- Launch a SageMaker Studio domain along with a *studio-user*
- Create a standard EMR cluster with no authentication enabled, including the other infrastructure required, such as a **Virtual Private Cloud** (**VPC**), subnets, and other resources

> **Note**
> You will incur a cost for EMR when you launch this CloudFormation template. Therefore, make sure to refer to the *Clean up* section at the end of the chapter.

The CloudFormation template that we will use in the book is originally taken from `https://aws-ml-blog.s3.amazonaws.com/artifacts/sma-milestone1/template_no_auth.yaml` and has been modified to run the code provided with the book.

To get started with launching the CloudFormation template, use your AWS account to run the following steps:

1. Log into your AWS account and open the SageMaker management console (`https://console.aws.amazon.com/sagemaker/home`), preferably as an admin user. If you don't have admin user access, make sure you have permission to create an EMR cluster, Amazon SageMaker Studio, and S3. You can refer to *Required Permissions* (`https://docs.aws.amazon.com/sagemaker/latest/dg/studio-notebooks-emr-required-permissions.html`) for details on the permission needed.

2. Go to the S3 bucket and upload the contents of the S3 folder from the GitHub repository. Go to the `templates` folder, click on `template_no_auth.yaml`, and copy `Object URL`.

3. Make sure you have the `artifacts` folder parallel to the `templates` folder in the S3 bucket as well.

4. Search for the `CloudFormation` service and click on it.

5. Once in the CloudFormation console, click on the **Create stack** orange button, as shown in *Figure 5.1*:

Figure 5.1 – AWS CloudFormation console

6. In the **Specify template** section, select **Amazon S3 URL** as **Template source** and enter **Amazon S3 URL** noted in *step 2*, as shown in *Figure 5.2*, and click on the **Next** button:

Create stack
Prerequisite - Prepare template
Prepare template
Every stack is based on a template. A template is a JSON or YAML file that contains configuration information about the AWS resources you want to include in the stack.
Template is ready
Use a sample template
Create template in Designer
Specify template
A template is a JSON or YAML file that describes your stack's resources and properties.
Template source
Selecting a template generates an Amazon S3 URL where it will be stored.
Amazon S3 URL
Upload a template file
Amazon S3 URL
/template/template_no_auth.yaml
Amazon S3 template URL
/template/template_no_auth.yaml
View in Designer
Cancel
Next

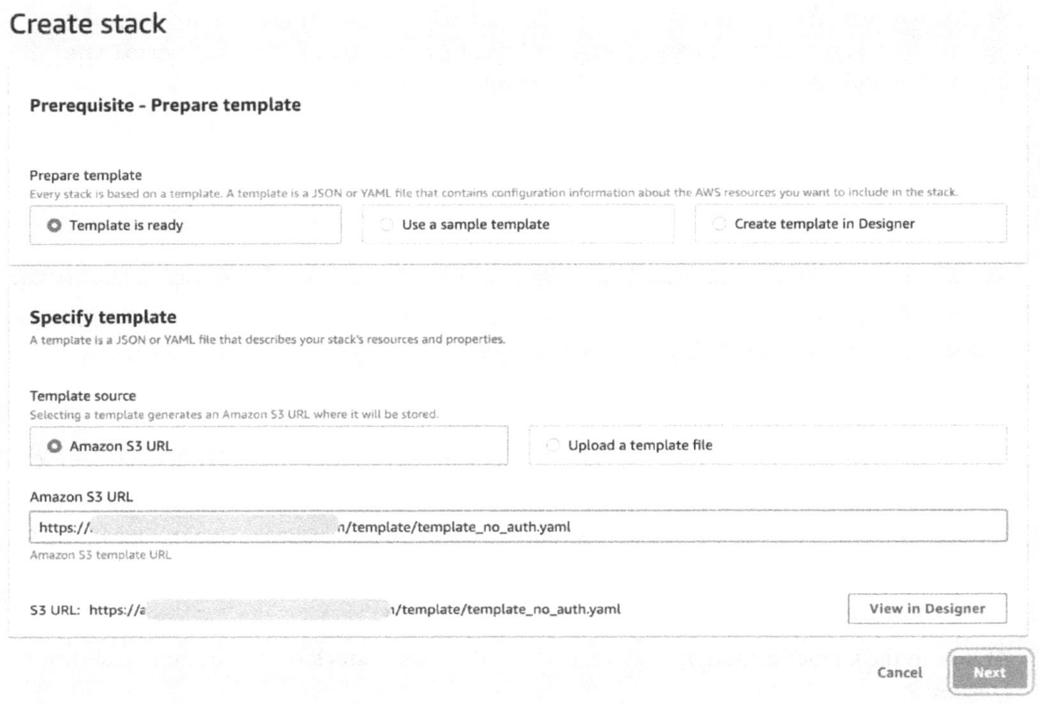

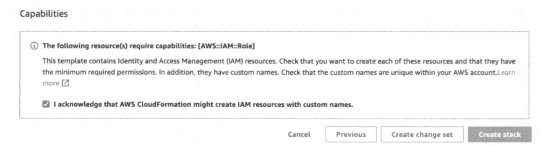

Figure 5.2 – Create stack

7. Enter the stack name of your choice and click on the **Next** button.
8. On the **Configure stack options** page, keep the default settings and click on the **Next** button at the bottom of the page.
9. On the **Review** page, scroll to the bottom of the screen, click on the **I acknowledge that AWS CloudFormation might create IAM resources with custom names** checkbox, and click on the **Create stack** button, as shown in *Figure 5.3*:

Capabilities
The following resource(s) require capabilities: [AWS::IAM::Role]
This template contains Identity and Access Management (IAM) resources. Check that you want to create each of these resources and that they have the minimum required permissions. In addition, they have custom names. Check that the custom names are unique within your AWS account.Learn more
I acknowledge that AWS CloudFormation might create IAM resources with custom names.
Cancel
Previous
Create change set
Create stack

Figure 5.3 – Review stack

> **Note**
>
> The CloudFormation template will take 5-10 minutes to launch.

10. Once it is launched, go to **Amazon SageMaker**, click on **SageMaker Studio**, and you will see **SageMaker Domain** and **studio-user** configured for you, as shown in *Figure 5.4*:

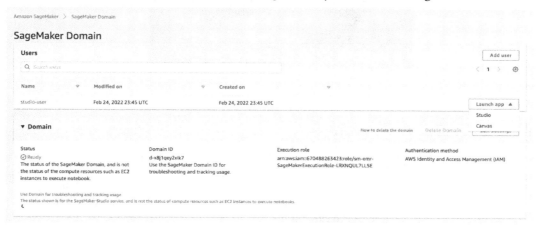

Figure 5.4 – SageMaker Domain

11. Click on the **Launch app** dropdown next to **studio-user** and select **Studio**, as shown in the preceding screenshot. After this, you will be presented with a new JupyterLab interface (https://jupyterlab.readthedocs.io/en/latest/user/interface.html).

> **Note**
>
> It is recommended that you familiarize yourself with the Studio UI by reviewing the Amazon SageMaker Studio UI documentation (https://docs.aws.amazon.com/sagemaker/latest/dg/studio-ui.html), as we will be referencing many of the SageMaker-specific widgets and views throughout the chapter.

12. To make it easier to run the various examples within the book using the Studio UI, we will clone the source code from the companion GitHub repository. Within the Studio UI, click on the **Git** icon in the left sidebar and once the resource panel opens, click on the **Clone a repository** button to launch the **Clone a repo** dialog box, as shown in *Figure 5.5*:

Figure 5.5 – Clone a repo

13. Enter the URL for the companion repository (`https://github.com/PacktPublishing/Applied-Machine-Learning-and-High-Performance-Computing-on-AWS.git`) and click the **CLONE** button.

14. The cloned repository will now appear in the **File Browser** panel of the Studio UI. Double-click on the newly cloned `Applied-Machine-Learning-and-High-Performance-Computing-on-AWS` folder to expand it.

15. Then double-click on the `Chapter05` folder to open it for browsing.

We are now ready to analyze large amounts of structured data using SageMaker Studio. However, before we can start the analysis, we need to acquire the data. Let's take a look at how to do that.

Analyzing large amounts of structured data

Since the objective of this section is to provide a hands-on example for analyzing large-scale structured data, our first task will be to synthesize a large amount of data. Using the Studio UI, execute the following steps:

1. Using the left **File Browser** panel, double-click on the `1_data_generator.ipynb` file to launch the Jupyter notebook.

2. When prompted with the **Set up notebook environment** dialog box, ensure that **Data Science** is selected from the **Image** drop-down box, as well as **Python 3** for the **Kernel** option. *Figure 5.6* shows an example of the dialog box:

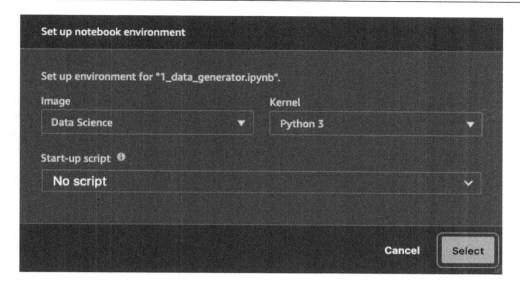

Figure 5.6 – Set up notebook environment

3. Once these options have been set, click the **Select** button to continue.

4. Next, you should see a **Starting notebook kernel...** message. The notebook kernel will take a couple of minutes to load.

5. Once the notebook has loaded, run the notebook by clicking on the **Kernel** menu and selecting the **Restart kernel and Run All Cells...** option.

After the notebook has executed all the code cells, we can dive into exactly what the notebook does, starting with a review of the dataset.

Reviewing the dataset

The dataset we will be using within this example is the California housing dataset (https://www.dcc.fc.up.pt/~ltorgo/Regression/cal_housing.html). This dataset was derived from the 1990 US census, using one row per census block group. A block group is the smallest geographical unit for which the US Census Bureau publishes sample data. A block group typically has a population of 600 to 3,000 people.

The dataset is incorporated into the **scikit-learn** or sklearn Python library (https://scikit-learn.org/stable/index.html). The scikit-learn library includes a dataset module that allows us to download popular reference datasets, such as the California housing dataset, from **StatLib Datasets Archive** (http://lib.stat.cmu.edu/datasets/).

> **Dataset citation**
>
> Pace, R. Kelley, and Ronald Barry, Sparse Spatial Autoregressions, Statistics and Probability Letters, 33 (1997) 291-297.

One key thing to be aware of is that this dataset only has 20,640 samples and is only around 400 KB in size. So, I'm sure you'll agree that it doesn't exactly qualify as a large amount of structured data. So, the primary objective of the notebook we've just executed is to use this dataset as a basis for synthesizing a much larger amount of structured data and then storing this new dataset on S3 for analysis.

Let's walk through the code to see how this is done.

Installing the Python libraries

The first five code cells within the notebook are used to install and upgrade the necessary Python libraries to ensure we have the correct versions for the SageMaker SDK, scikit-learn, and the **Synthetic Data Vault**. The following code snippet shows the consolidation of these five code cells:

```
. . .
import sys
!{sys.executable} -m pip install "sagemaker>=2.51.0"
!{sys.executable} -m pip install --upgrade -q "scikit-learn"
!{sys.executable} -m pip install "sdv"

import sklearn
sklearn.__version__

import sdv
sdv.__version__
. . .
```

> **Note**
>
> There is no specific reason we upgrade and install the SageMaker and scikit-learn libraries except to ensure conformity across the examples within this chapter.

Once the required libraries have been installed, we load them and configure our global variables. The following code snippet shows how we import the libraries and configure the SageMaker default S3 bucket parameters:

```
...
import os
from sklearn.datasets import fetch_california_housing
import time
import boto3
import numpy as np
import pandas as pd
from sklearn.model_selection import train_test_split
import sagemaker
from sagemaker import get_execution_role

prefix = 'california_housing'
role = get_execution_role()
bucket = sagemaker.Session(boto3.Session()).default_bucket()
...
```

However, before we can synthesize a larger dataset and upload this to S3, we need to download the California housing dataset. As you can see from the following code snippet, we create two local folders called `data` and `raw`, then download the data using the `fetch_california_housing()` method from `sklearn.datasets`. The resultant data variable allows us to describe the data, as well as capture the data itself as a two-dimensional data structure called `df_data`:

```
...
data_dir = os.path.join(os.getcwd(), "data")
os.makedirs(data_dir, exist_ok=True)

raw_dir = os.path.join(os.getcwd(), "data/raw")
os.makedirs(raw_dir, exist_ok=True)

data = fetch_california_housing(data_home=raw_dir, download_if_
missing=True, return_X_y=False, as_frame=True)
...
df_data = data.data
...
```

The df_data variable is the essential representation of our structured data, with columns showing the data labels and rows showing the observations or records for each label. Think of this structure as similar to a spreadsheet or relational table.

Using the df_data variable, we further describe this structure as well as perform some of the descriptive statistics and visualization described in the *Exploring data analysis methods* section of this chapter. For example, the following code snippet shows how to describe the type of data we are dealing with. You will recall that understanding the data type is crucial for appreciating the overall schema or structure of the data:

```
. . .
df_data.astype({'Population': 'int32'}).dtypes
. . .
```

Furthermore, we can define a Python function called plot_boxplot() to visualize the data included in the df_data variable. You will recall that visualizing the data provides further insight into the data. For example, as you can see from the next code snippet, we can visualize the overall distribution of the average number of rooms or avgNumrooms in the house:

```
. . .
import matplotlib.pyplot as plt

def plot_boxplot(data, title):
    plt.figure(figsize =(5, 4))
    plt.boxplot(data)
    plt.title(title)
    plt.show()
. . .
df_data.drop(df_data[df_data['avgNumRooms'] > 9].index, inplace
= True)
df_data.drop(df_data[df_data['avgNumRooms'] <= 1].index,
inplace = True)
plot_boxplot(df_data.avgNumRooms, 'rooms')
. . .
```

As we can see from *Figure 5.7*, the resultant boxplot from the code indicates that the average number of rooms for the California housing data is **5**:

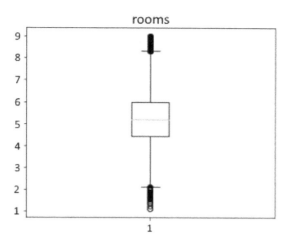

Figure 5.7 – Average number of rooms

Additionally, you will note from *Figure 5.7* that there is somewhat of an equal distribution to the upper and lower bounds of the data. This indicates that we have a good distribution of data for the average number of bedrooms and therefore, we don't need to augment this data point.

Finally, you will recall from the *Counting the data points* section that we can circumvent any unnecessary issues during the model training process by determining whether or not there are any missing values in the data. For example, the next code snippet shows how we can review a sum of any missing values in the df_data variable:

```
. . .
df_data.isna().sum()
. . .
```

While we've only covered a few analytics methodologies to showcase the analytics life cycle, a key takeaway from these examples is that the data is easy to analyze since it's small enough to fit into memory. So, as data scientists, we did not have to capture a sample of the global population to analyze the data and then infer that analysis back onto the larger dataset. Let's see whether this holds true once we synthesize a larger dataset.

Synthesizing large data

The last part of the notebook involves using the Synthetic Data Vault (https://sdv.dev/SDV/index.html), or the sdv Python library. This ecosystem of Python libraries uses ML models that specifically focus on learning from structured tabular and time series datasets and on creating synthetic data that carries the same format, statistical properties, and structure as the original dataset.

In our example notebook, we use a TVAE (https://arxiv.org/abs/1907.00503) model to generate a larger version of the California housing data. For example, the following code snippet shows how we define and train a TVAE model on the df_data variable:

```
...
from sdv.tabular import TVAE
model = TVAE(rounding=2)
model.fit(df_data)

model_dir = os.path.join(os.getcwd(), "model")
os.makedirs(model_dir, exist_ok=True)

model.save(f'{model_dir}/tvae_model.pkl')
...
```

Once we've trained the model, we can load it to generate 1 million new observations or rows in a variable called synthetic_data. The following code snippet shows an example of this:

```
...
from sdv.tabular import TVAE
model = TVAE.load(f'{model_dir}/tvae_model.pkl')
synthetic_data = model.sample(1000000)
...
```

Finally, as shown, we use the following code snippet to compress the data and leverage the SageMaker SDK's upload_data() method to store the data in S3:

```
...
sess = boto3.Session()
sagemaker_session = sagemaker.Session(boto_session=sess)

synthetic_data.to_parquet('data/raw/data.parquet.gzip',
compression='gzip')

rawdata_s3_prefix = "{}/data/raw".format(prefix)
raw_s3 = sagemaker_session.upload_data(path="./data/raw/data.
parquet.gzip", key_prefix=rawdata_s3_prefix)
...
```

With a dataset of 1 million rows stored in S3, we finally have an example of a large amount of structured data. Now we can use this data to demonstrate how to leverage the highlighted analysis methods at scale on structured data using Amazon EMR.

Analyzing large-scale data using an EMR cluster with SageMaker Studio

To get started with analyzing the large-scale synthesized dataset we've just created, we can execute the following steps in the Studio UI:

1. Using the left-hand navigation panel, double-click on the 2_data_exploration_spark. ipynb notebook to launch it.
2. As we saw with the previous example, when prompted with the **Set up notebook environment** dialog box, select **SparkMagic** as **Image** and **PySpark** as **Kernel**.
3. Once the notebook is ready, click on the **Kernel** menu option and once again select the **Restart kernel and Run All Cells…** option to execute the entire notebook.

While the notebook is running, we can start reviewing what we are trying to accomplish within the various code cells. As you can see from the first code cell, we connect to the EMR cluster we provisioned in the *Setting up EMR and SageMaker Studio* section:

```
%load_ext sagemaker_studio_analytics_extension.magics
%sm_analytics emr connect --cluster-id <EMR Cluster ID> --auth-
type None
```

In the next code cell, shown by the following code, we read the synthesized dataset from S3. Here we create a housing_data variable by using PySpark's sqlContext method to read the raw data from S3:

```
housing_data=sqlContext.read.parquet('s3://<SageMaker Default
Bucket>/california_housing/data/raw/data.parquet.gzip')
```

Once we have this variable assigned, we can use PySpark and the EMR cluster to execute the various data analysis tasks on the entire population of the data without having to ingest a sample dataset on which to perform the analysis.

While the notebook provides multiple examples of different analysis methodologies that are specific to the data, we will focus on the few examples that relate to the exploration we've already performed on the original California housing dataset to illustrate how these same methodologies can be applied at scale.

Reviewing the data structure and counts

As already mentioned, understanding the type of data, its structure, and the counts is an important part of the analysis. To perform this analysis on the entirety of housing_data, we can execute the following code:

```
print((housing_data.count(), len(housing_data.columns)))
housing_data.printSchema()
```

Executing this code produces the following output, where we can see that we have 1 million observations, as well as the data types for each feature:

```
(1000000, 9)
Root
 |-- medianIncome: double (nullable = true)
 |-- medianHousingAge: double (nullable = true)
 |-- avgNumRooms: double (nullable = true)
 |-- avgNumBedrooms: double (nullable = true)
 |-- population: double (nullable = true)
 |-- avgHouseholdMembers: double (nullable = true)
 |-- latitude: double (nullable = true)
 |-- longitude: double (nullable = true)
 |-- medianHouseValue: double (nullable = true)
```

Next, we can determine whether or not there are any missing values and how to deal with them.

Handling missing values

You will recall that ensuring that there are no missing values is an important methodology for any data analysis. To expose any missing data, we can run the following code to create a count of any missing values for each column or feature within this large dataset:

```
from pyspark.sql.functions import isnan, when, count, col
housing_data.select([count(when(isnan(c), c)).alias(c) for c in
housing_data.columns]).show()
```

If we do find any missing values, there are a number of techniques we can use to deal with them. For example, we delete rows containing missing values, using the dropna() method on the housing_data variable. Alternatively, depending on the number of missing values, we can use imputation techniques to infer a value based on the mean or median of the feature.

Analyzing the centrality and variability of the data

Remember that understanding how the data is distributed will give us a good idea of whether the data is proportional or not. This analysis task also provides an idea of whether we have outliers within our data that skew the distribution or spread. Previously, it was emphasized that visualizing the data distribution using bar charts and boxplots can further assist in determining the variability of the data.

To accommodate this task, the following code highlights an example of capturing the features we wish to analyze and plotting their distribution as a bar chart:

```
import matplotlib.pyplot as plt

df = housing_data.select('avgNumRooms', 'avgNumBedrooms',
'population').toPandas()
df.hist(figsize=(10, 8), bins=20, edgecolor="black")
plt.subplots_adjust(hspace=0.3, wspace=0.5)
plt.show()
%matplot plt
```

After executing this code on the large data, we can see an example of the resultant distribution for the average number of rooms (`avgNumRooms`), the average number of bedrooms (`avgNumBedrooms`), and block population (`population`) features in *Figure 5.8*:

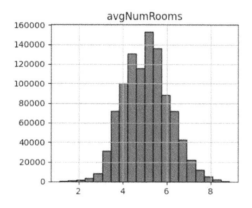
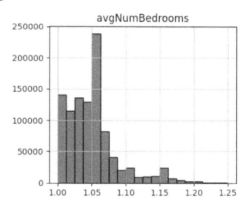
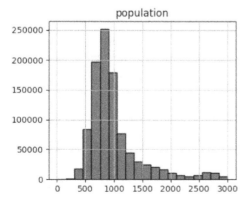

Figure 5.8 – Feature distribution

As you can see from *Figure 5.8*, both the `avgNumBedrooms` and `population` features are not centered around the mean or average for the feature. Additionally, the spread for the `avgNumBedrooms` feature is significantly skewed toward the lower end of the spectrum. This factor could indicate that there are potential outliers or too many data points that are consolidated between **1.00** and **1.05**. This fact is further confirmed if we use the following code to create a boxplot of the `avgNumBedrooms` feature:

```
plot_boxplot(df.avgNumBedrooms, 'Boxplot for Average Number of
Bedrooms')
%matplot plt
```

The resultant boxplot from running this code cell is shown in *Figure 5.9*:

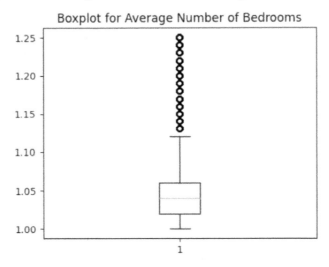

Figure 5.9 – Boxplot for the average number of bedrooms

Figure 5.9 clearly shows that there are a number of outliers that cause the data to be skewed. Therefore, we need to resolve these discrepancies as part of our data analysis and before ML models can be trained on our large dataset. The following code snippet shows how to query the data from the boxplot values and then simply remove it, to create a variable called `housing_df_with_no_outliers`:

```
import pyspark.sql.functions as f

columns = ['avgNumRooms', 'avgNumBedrooms', 'population']

housing_df_with_no_outliers = housing_data.where(
    (housing_data.avgNumRooms<= 8) &
    (housing_data.avgNumRooms>=2) &
```

```
    (housing_data.avgNumBedrooms<=1.12) &
    (housing_data.population<=1500) &
    (housing_data.population>=250))
```

Once we have our housing_df_with_no_outliers, we can use the following code to create a new boxplot of the variability of the avgNumBedrooms feature:

```
df = housing_df_with_no_outliers.select('avgNumRooms',
'avgNumBedrooms', 'population').toPandas()
plot_boxplot(df.avgNumBedrooms, 'Boxplot for Average Number of
Bedrooms')
%matplot plt
```

Figure 5.10 shows an example of a boxplot produced from executing this code:

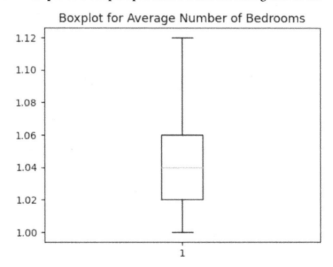

Figure 5.10 – Boxplot of the average number of bedrooms

From *Figure 5.10*, we can clearly see that the outliers have been removed. Subsequently, we can perform a similar procedure on the avgNumRooms and population features.

While these examples only show some of the important methodologies highlighted within the data analysis life cycle, an important takeaway from this exercise is that due to the integration of SageMaker Studio and EMR, we're able to accomplish the data analysis tasks on large-scale structured data without having to capture a sample of the global population and then infer that analysis back onto the larger dataset. However, along with analyzing the data at scale, we also need to ensure that any preprocessing tasks are also executed at scale.

Next, we will review how to automate these preprocessing tasks at scale using SageMaker.

Preprocessing the data at scale

The SageMaker service takes care of the heavy lifting and scaling of data transformation or preprocessing tasks using one of its core components called **Processing jobs** (https://docs.aws.amazon.com/sagemaker/latest/dg/processing-job.html). While Processing jobs allows a user to leverage built-in images for scikit-learn or even custom images, they also reduce the heavy-lifting task of provisioning ephemeral Spark clusters (https://docs.aws.amazon.com/sagemaker/latest/dg/use-spark-processing-container.html). This means that data scientists can perform large-scale data transformations automatically without having to create or have the infrastructure team create an EMR cluster. All that's required of the data scientist is to convert the processing code cell into a Python script called preprocess.py.

The following code snippet shows how the code to remove outliers can be converted into a Python script:

```python
%%writefile preprocess.py
...
def main():
    parser = argparse.ArgumentParser(description="app inputs
and outputs")
    parser.add_argument("--bucket", type=str, help="s3 input
bucket")
    parser.add_argument("--s3_input_prefix", type=str, help="s3
input key prefix")
    parser.add_argument("--s3_output_prefix", type=str,
help="s3 output key prefix")
    args = parser.parse_args()
    spark = SparkSession.builder.appName("PySparkApp").
getOrCreate()
    housing_data=spark.read.parquet(f's3://{args.bucket}/{args.
s3_input_prefix}/data.parquet.gzip')
    housing_df_with_no_outliers = housing_data.where((housing_
data.avgNumRooms<= 8) &
                    (housing_data.avgNumRooms>=2) &
                    (housing_data.avgNumBedrooms<=1.12) &
                    (housing_data.population<=1500) &
                    (housing_data.population>=250))
    (train_df, validation_df) = housing_df_with_no_outliers.
randomSplit([0.8, 0.2])
    train_df.write.parquet("s3://" + os.path.join(args.bucket,
```

```
args.s3_output_prefix, "train/"))
    validation_df.write.parquet("s3://" + os.path.join(args.
bucket, args.s3_output_prefix, "validation/"))


if __name__ == "__main__":
    main()
...
```

Once the Python script is created, we can load the appropriate SageMaker SDK libraries and configure the S3 locations for the input data as well as the transformed output data, as follows:

```
%local
import sagemaker
from time import gmtime, strftime
sagemaker_session = sagemaker.Session()
role = sagemaker.get_execution_role()
bucket = sagemaker_session.default_bucket()
timestamp = strftime("%Y-%m-%d-%H-%M-%S", gmtime())
prefix = "california_housing/data_" + timestamp
s3_input_prefix = "california_housing/data/raw"
s3_output_prefix = prefix + "/data/spark/processed"
```

Finally, we can instantiate an instance of the SageMaker PySparkProcessor() class (https://sagemaker.readthedocs.io/en/stable/api/training/processing.html#sagemaker.spark.processing.PySparkProcessor) as a spark_processor variable, as can be seen in the following code:

```
%local
from sagemaker.spark.processing import PySparkProcessor
spark_processor = PySparkProcessor(
    base_job_name="sm-spark",
    framework_version="2.4",
    role=role,
    instance_count=2,
    instance_type="ml.m5.xlarge",
    max_runtime_in_seconds=1200,
)
```

With the `spark_processor` variable defined, we can then call the `run()` method to execute a SageMaker Processing job. The following code demonstrates how to call the `run()` method and supply the `preprocess.py` script along with the input and output locations for the data as `arguments`:

```
spark_processor.run(
    submit_app="preprocess.py",
    arguments=[
        "--bucket",
        bucket,
        "--s3_input_prefix",
        s3_input_prefix,
        "--s3_output_prefix",
        s3_output_prefix,
    ],
)
```

In the background, SageMaker will create an ephemeral Spark cluster and execute the `preprocess.py` script on the input data. Once the data transformations are completed, SageMaker will store the resultant dataset on S3 and then decommission the Spark cluster, all while redirecting the execution log output back to the Jupyter notebook.

While this technique makes the complicated task of analyzing large amounts of structured data much easier to scale, there is still the question of how to perform a similar procedure on unstructured data.

Let's review how to solve this problem next.

Analyzing large amounts of unstructured data

In this section, we will use unstructured data (horse and human images) downloaded from https://laurencemoroney.com/datasets.html. This dataset can be used to train a binary image classification model to classify horses and humans in the image. From the SageMaker Studio, launch the `3_unstructured_data_s3.ipynb` notebook with **PyTorch 1.8 Python 3.6 CPU Optimized** selected from the **Image** drop-down box as well as **Python 3** for the **Kernel** option. Once the notebook is opened, restart the kernel and run all the cells as mentioned in the *Analyzing large-scale data using an EMR cluster with SageMaker Studio* section.

After the notebook has executed all the code cells, we can dive into exactly what the notebook does.

As you can see in the notebook, we first download the horse-or-human data from https://storage.googleapis.com/laurencemoroney-blog.appspot.com/horse-or-human.zip and then unzip the file.

Once we have the data, we will convert the images to high resolution using the **EDSR** model provided by the **Hugging Face** `super-image` library:

1. We will first download the pretrained model with `scale = 4`, which means that we intend to increase the resolution of the images four times, as shown in the following code block:

    ```
    from super_image import EdsrModel, ImageLoader
    from PIL import Image
    import requests
    model = EdsrModel.from_pretrained('eugenesiow/edsr-base',
    scale=4)
    ```

2. Next, we will iterate through the folder containing the images, use the pretrained model to convert each image to high resolution, and save it, as shown in the following code block:

    ```
    import os
    from os import listdir
    folder_dir = "horse-or-human/"
    for folder in os.listdir(folder_dir):
        folder_path = f'{folder_dir}{folder}'
        for image_file in os.listdir(folder_path):
            path = f'{folder_path}/{image_file}'
            image = Image.open(path)
            inputs = ImageLoader.load_image(image)
            preds = model(inputs)
            ImageLoader.save_image(preds, path)
    ```

You can check the file size of one of the images to confirm that the images have been converted to high resolution. Once the images have been converted to high resolution, you can optionally duplicate the images to increase the number of files and finally upload them to S3 bucket. We will use the images uploaded to the S3 bucket for running a SageMaker training job.

Note

In this example, we will walk you through the option of running a training job with **PyTorch** using the SageMaker training feature using the data stored in an S3 bucket. You can also choose to use other frameworks as well, such as **TensorFlow** and **MXNet**, which are also supported by SageMaker.

In order to use PyTorch, we will first import the `sagemaker.pytorch` module, using which we will define the **SageMaker PyTorch Estimator**, as shown in the following code block:

```
from sagemaker.pytorch import PyTorch
estimator = PyTorch(entry_point='train.py',
                    source_dir='src',
                    role=role,
                    instance_count=1,
                    instance_type='ml.g4dn.8xlarge',
                    framework_version='1.8.0',
                    py_version='py3',
                    sagemaker_session=sagemaker_session,
                    hyperparameters={'epochs':5,
                                     'subset':2100,
                                     'num_workers':4,
                                     'batch_size':500},
                    )
```

In the estimator object, as you can see from the code snippet, we need to provide configuration parameters. In this case, we need to define the following parameters:

- `instance_count`: This is the number of instances
- `instance_type`: This is the type of instance on which the training job will be launched
- `framework_version`: This is the framework version of PyTorch to be used for training
- `py_version`: This is the Python version
- `source_dir`: This is the folder path within the notebook, which contains the training script
- `entry_point`: This is the name of the Python training script
- `hyperparameters`: This is the list of hyperparameters that will be used by the training script

Once we have defined the PyTorch estimator, we will define the `TrainingInput` object, which will take the S3 location of the input data, content type, and input mode as parameters, as shown in the following code:

```
from sagemaker.inputs import TrainingInput
train = TrainingInput(s3_input_data,content_type='image/
png',input_mode='File')
```

The `input_mode` parameter can take the following values; in our case, we are using the `File` value:

- `None`: Amazon SageMaker will use the input mode specified in the base `Estimator` class

- `File`: Amazon SageMaker copies the training dataset from the S3 location to a local directory

- `Pipe`: Amazon SageMaker streams data directly from S3 to the container via a *Unix-named pipe*

- `FastFile`: Amazon SageMaker streams data from S3 on demand instead of downloading the entire dataset before training begins

> **Note**
>
> You can see the complete list of parameters for the PyTorch estimator at this link: `https://sagemaker.readthedocs.io/en/stable/frameworks/pytorch/sagemaker.pytorch.html`.

Once we have configured the PyTorch `Estimator` and `TrainingInput` objects, we are now ready to start the training job, as shown in the following code:

```
estimator.fit({'train':train})
```

When we run `estimator.fit`, it will launch one training instance of the `ml.g4dn.8xlarge` type, install the `PyTorch 1.8` container, copy the training data from `S3 location` and the `train.py` script to the local directory on the training instance, and will finally run the training script that you have provided in the estimator configuration. Once the training job is completed, SageMaker will automatically terminate all the resources that it has launched, and you will only be charged for the amount of time the training job was running.

In this example, we used a simple training script that involved loading the data using the PyTorch `DataLoader` object and iterating through the images. In the following section, we'll see how to process data at scale using AWS.

Processing data at scale on AWS

In the previous section, *Analyzing large amounts of unstructured data*, the data was stored in an S3 bucket, which was used for training. There will be scenarios where you will need to load data faster for training instead of waiting for the training job to copy the data from S3 locally into your training instance. In these scenarios, you can store the data on a file system, such as **Amazon Elastic File System** (**EFS**) or **Amazon FSx,** and mount it to the training instance, which will be faster than storing the data in S3 location. The code for this is in the `3_unstructured_data.ipynb` notebook. Refer to the **Optimize it with data on EFS** and **Optimize it with data on FSX** sections in the notebook.

> **Note**
>
> Before you run the **Optimize it with data on EFS** and **Optimize it with data on FSX** sections, please launch the CloudFormation `template_filesystems.yaml` template, in a similar fashion as we did in the *Setting up EMR and SageMaker Studio* section.

Cleaning up

Let's terminate the EMR cluster, which we launched in the *Setting up EMR and SageMaker Studio* section, as it will not be used in the later chapters of the book.

Let's start by logging into the AWS console and following the steps given here:

1. Search EMR in the AWS console.

2. You will see the active **EMR-Cluster-sm-emr** cluster. Select the checkbox against the EMR cluster name and click on the **Terminate** button, as shown in *Figure 5.11*:

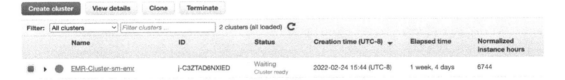

Figure 5.11 – List EMR cluster

3. Click on the red **Terminate** button in the pop-up window, as shown in *Figure 5.12*:

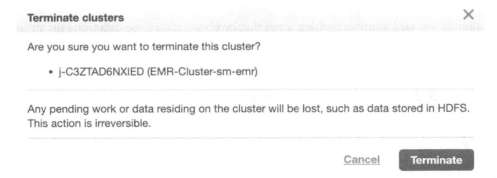

Figure 5.12 – Terminate EMR cluster

4. It will take a few minutes to terminate the EMR cluster, and once completed, **Status** will change to **Terminated**.

Let's summarize what we've learned in this chapter.

Summary

In this chapter, we explored various data analysis methods, reviewed some of the AWS services for analyzing data, and launched a CloudFormation template to create an EMR cluster, SageMaker Studio domain, and other useful resources. We then did a deep dive into code for analyzing both structured and unstructured data and suggested a few methods for optimizing its performance. This will help you to prepare your data for training ML models.

In the next chapter, we will see how we can train large models on large amounts of data in a distributed fashion to speed up the training process.

6

Distributed Training of Machine Learning Models

When it comes to **Machine Learning (ML)** model training, the primary goal for a data scientist or ML practitioner is to train the optimal model based on the relevant data to address the business use case. While this goal is of primary importance, the panacea is to perform this task as quickly and effectively as possible. So, *how do we speed up model training?* Moreover, sometimes, the data or the model might be too big to fit into a single GPU memory. *So how do we prevent out-of-memory (OOM) errors?*

The simplest answer to this question is to basically throw more compute resources, in other words, more CPUs and GPUs, at the problem. This is essentially using larger compute hardware and is commonly referred to as a **scale-up** strategy. However, there is only a finite number of CPUs and GPUs that can be squeezed into a server. So, sometimes a **scale-out** strategy is required, whereby we add more servers into the mix, essentially distributing the workload across multiple physical compute resources.

Nonetheless, spreading the model training workload across more CPUs or GPUs, and even across more compute servers, will definitely speed up the overall training process. Making use of either a scale-up, scale-out, or a combination of the two strategies also adds further complexity to the overall orchestration and configuration of the model training activity. Therefore, this chapter will help navigate these challenges to help overcome the additional complexities imposed by the **distributed training** process by covering the following topics:

- Building ML systems using AWS
- Introducing the fundamentals of distributed training
- Executing a distributed training workload on AWS

Technical requirements

You should have the following prerequisites before getting started with this chapter:

- A web browser (for the best experience, it is recommended that you use a Chrome or Firefox browser)

- Access to the AWS account that you used in *Chapter 5, Data Analysis*

- An AWS account (if you are unfamiliar with how to get started with an AWS account, you can go to this link `https://aws.amazon.com/getting-started/`)

- Access to the SageMaker Studio development environment that we created in *Chapter 5, Data Analysis*

- Example Jupyter notebooks for this chapter are provided in the companion GitHub repository (`https://github.com/PacktPublishing/Applied-Machine-Learning-and-High-Performance-Computing-on-AWS/tree/main/Chapter06`)

Building ML systems using AWS

Before we can explore the fundamentals of how to implement the distributed training strategies highlighted at the outset, we first need to level set and understand just how the ML model training exercise can be performed on the AWS platform. Once we understand how AWS handles model training, we can further expand on this concept to address the concept of distributed training.

To assist ML practitioners in building ML systems, AWS provides the SageMaker (`https://aws.amazon.com/sagemaker/`) service. While SageMaker is a single AWS service, it comprises multiple modules that map specifically to an ML task. For example, SageMaker provides the Training job component that is purpose-built to take care of the heavy lifting and scaling of the model training task. ML practitioners can use SageMaker Training jobs to essentially provision ephemeral compute environments or clusters to handle the model training task. Essentially, all the ML practitioner needs to do is specify a few configuration parameters, and SageMaker Training jobs takes care of the rest. For example, we need to supply the following four basic parameters:

- The URL for the S3 bucket, which contains the model training, testing, and optionally, the validation data

- The type and quantity of ML compute instances required to perform the model training task

- The location of the S3 bucket to store the trained model

- The location, either locally or on S3, where the model training code is stored

The following code snippet shows just how easy it can be to formalize these four basic requirements into a SageMaker Training job request:

```
. . .
from sagemaker.pytorch import PyTorch
estimator = PyTorch(entry_point='train.py',
                    source_dir='src',
                    role=role,
                    instance_count=1,
                    instance_type='ml.p3.2xlarge',
                    framework_version='1.8.0',
                    py_version='py3',
                    sagemaker_session=sagemaker_session,
                    hyperparameters={'epochs':10,
                                     'batch_size':32,
                                     'lr':3e-5,
                                     'gamma': 0.7},
                    )
. . .
```

Using this code snippet, we basically tell SageMaker that we want to use the built-in PyTorch estimator by declaring the `estimator` variable to use the PyTorch framework. We then supply the necessary requirements, such as the following:

- `entry_point`: This is the location of the training script.
- `instance_count`: This is the number of compute servers to be provisioned in the cluster.
- `instance_type`: This is the type of compute resources required in the cluster. In this example, we are specifying the `ml.p3.16xlarge` instances.

> **Note**
>
> For more information on the SageMaker PyTorch estimator, as well as how to leverage the SageMaker SDK to instantiate the estimator, see the AWS documentation on how to use PyTorch on SageMaker (`https://docs.aws.amazon.com/sagemaker/latest/dg/pytorch.html`) and the SageMaker SDK documentation (`https://sagemaker.readthedocs.io/en/stable/frameworks/pytorch/sagemaker.pytorch.html#pytorch-estimator`).

Once we have declared the estimator, we specify the location of the training and validation datasets on S3, as shown in the following code snippet:

```
...
from sagemaker.inputs import TrainingInput

train = TrainingInput(s3_train_data,
                      content_type='image/png',
                      input_mode='File')
val = TrainingInput(s3_val_data,
                    content_type='image/png',
                    input_mode='File')
...
```

We then call the `fit()` method of the PyTorch estimator to tell SageMaker to execute the Training job on the datasets, as shown in the following code snippet:

```
...
estimator.fit({'train':train, 'val': val})
...
```

Behind the scenes, SageMaker creates an ephemeral compute cluster, executes the training task on these resources, and then produces the resultant optimized model, which is then stored on Amazon S3. After this task has been performed, SageMaker tears down the ephemeral cluster with users only paying for the resources consumed during the training time.

> **Note**
>
> For more detailed information as to how SageMaker Training jobs work behind the scenes, see the AWS documentation (https://docs.aws.amazon.com/sagemaker/latest/dg/how-it-works-training.html).

So now that we have a basic idea of how a model training exercise can be performed using Amazon SageMaker, *how can we improve on model training time and essentially speed up the process by leveraging more compute resources?*

To answer this question, we can very easily implement a scale-up strategy with the SageMaker Training job. All we have to do is change the `instance_type` parameter for the `estimator` variable from `ml.p3.2xlarge` to `ml.p3.16xlarge`. By doing this, we are increasing the size of, or scaling up the compute resource from, an instance with 8 vCPUs, 61 GB of RAM, and a single GPU to an instance with 64 vCPUs, 488 GB of RAM, and 8 GPUs.

The resultant code now looks as follows:

```
...
from sagemaker.pytorch import PyTorch
estimator = PyTorch(entry_point='train.py',
                    source_dir='src',
                    role=role,
                    instance_count=1,
                    instance_type='ml.p3.16xlarge',
                    framework_version='1.8.0',
                    py_version='py3',
                    sagemaker_session=sagemaker_session,
                    hyperparameters={'epochs':10,
                                     'batch_size':32,
                                     'lr':3e-5,
                                     'gamma': 0.7},
                    )
...
```

So, as you can see, implementing a scale-up strategy is very straightforward when using SageMaker. However, *what if we need to go beyond the maximum capacity of an accelerated computing instance?*

Well, then, we would need to implement a scale-out strategy and distribute the training process across multiple compute nodes. In the next section, we will explore how to apply a scale-out strategy using distributed training for SageMaker Training jobs.

Introducing the fundamentals of distributed training

In the previous section, we highlighted how to apply a scale-up strategy to SageMaker Training jobs by simply specifying a large compute resource or large instance type. Implementing a scale-out strategy for the training process is just as straightforward. For example, we can increase the `instance_count` parameter for the Training job from 1 to 2 and thereby instruct SageMaker to instantiate an ephemeral cluster consisting of 2 compute resources as opposed to 1 node. Thus, the following code snippet highlights what the `estimator` variable configuration will look like:

```
...
from sagemaker.pytorch import PyTorch
estimator = PyTorch(entry_point='train.py',
                    source_dir='src',
                    role=role,
```

```
instance_count=2,
instance_type='ml.p3.2xlarge',
framework_version='1.8.0',
py_version='py3',
sagemaker_session=sagemaker_session,
hyperparameters={'epochs':10,
                 'batch_size':32,
                 'lr':3e-5,
                 'gamma': 0.7},

)
```

. . .

Unfortunately, simply changing the number of compute instances doesn't completely solve the problem. As already stated at the outset of this chapter, applying a scale-out strategy to distribute the SageMaker Training job adds further complexity to the overall orchestration and configuration of the model training activity. For example, when distributing the training activity, we need to also take into consideration the following aspects:

- *How do the various compute resources get access to and share the data?*

- *How do the compute resources communicate and coordinate their training tasks with each other?*

So, while simply specifying the number of compute resources for the Training job will create an appropriately sized training cluster, we also need to inform SageMaker about our **model placement strategy**. A model placement strategy directs SageMaker on exactly how the model is allocated or assigned to each of the compute resources within each node of the cluster. In turn, SageMaker uses the placement strategy to coordinate how each node interacts with the associated training data and, subsequently, how each node coordinates and communicates its portion of the model training task.

So how do we determine an effective model placement strategy for SageMaker?

The best way to answer this question is to understand what placement strategies are available and dissect how each of these strategies works. There are numerous placement strategies that are specific to each of the different training frameworks, as well as many open source frameworks. Nonetheless, all these different mechanisms can be grouped into two specific categories of placement strategies, namely **data parallel** and **model parallel**.

Let's explore SageMaker's data parallel strategy first.

Reviewing the SageMaker distributed data parallel strategy

As the name implies, a data parallel strategy focuses on the placement of model's training data. So, in order to fully understand just how this placement strategy is applied to the data, we should start by understanding just how a training activity interacts with the training data to optimize an ML model.

When we train an ML model, we basically create a training loop that applies the specific ML algorithm to the data. Typically, as is the case with deep learning algorithms, we break the data into smaller groups of records or batches of data. These batches are referred to as **mini batches**. We then pass each of these mini batches forward through the neural network layers, and then backward to optimize or train the model parameters. After completing one mini batch, we then apply the same procedure to the next mini batch, and so on, until we've run through the entirety of the data. A full execution of this process on the entire dataset is referred to as an **epoch**. Depending on the type of algorithm and, of course, the use case, we may have to run the algorithm to train the model for multiple epochs. It's this task that invariably takes the most time, and it's this task that we essentially want to improve on to reduce the overall time it takes to train the model.

So, when using a data parallel placement strategy, we are basically converting the training task from a sequential process to a parallel process. Instead of running the algorithm through a mini batch, then the next mini batch, then the next mini batch sequentially, we are now giving each individual mini batch to a separate compute resource, with each compute resource, in turn, running the model training process on its individual mini batch. Therefore, with each compute resource running its own mini batch at the same time, we are effectively distributing the epoch across multiple compute resources in parallel, therefore, improving the overall model training time. Consequently, using the data parallel technique does, however, introduce an additional complication, namely parallel optimization of all the weighted parameters for the model.

To further elaborate on this problem, we'll use the example depicted in *Figure 6.1*, detailing the individual node parameters for the model:

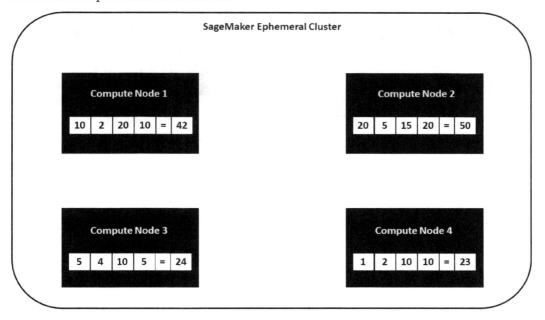

Figure 6.1 – Individual node parameters

As you can see from the example in *Figure 6.1*, we have four individual compute resources or nodes. Using the data parallel placement strategy, we have effectively placed a copy of the model algorithm onto each of these nodes and distributed the mini batch data across these resources. Now, each node computes the gradient reduction operation, in this case, the sum of weighted parameters of the model, on its individual mini batch of the data, in essence, producing four unique gradient calculation results. Since our goal is not to produce four separate representations of an optimized model but rather a single optimized model, *how do we combine the results across all four nodes?*

To solve this problem, SageMaker provides an additional optimization operation to the distributed training process and uses the **AllReduce** algorithm to share and communicate the results across the cluster. By including this additional step in the process, we can see the outcome in *Figure 6.2*:

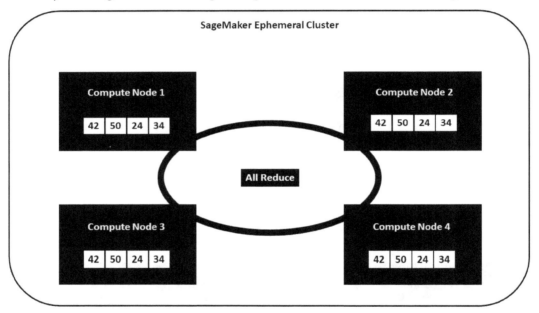

Figure 6.2 – Shared node parameters

From *Figure 6.2*, we can see that the AllReduce step takes the results from each node's gradient reduction operation and shares the results with every other node, ensuring that each node's representation of the model includes the optimizations from all the other nodes. This, therefore, guarantees that a single, consistent model is produced as the final output from the distributed training process.

> **Note**
>
> While the initial concept of using the AllReduce step for distributed deep learning was initially introduced in a blog post by Baidu Research, the original post has since been removed. So, for more background information on the intricacies of how it works, you can review the open source implementation called **Horovod** (https://eng.uber.com/horovod/).

Up until this point in the chapter, we have used the broad term *compute resources* to denote CPUs, GPUs, and physical compute instances. However, when implementing a successful data parallel placement strategy, it's important to fully understand just how SageMaker uses these compute resources to execute the distributed training workload.

Succinctly, when we instruct SageMaker to implement a data parallel placement strategy for the Training job, we are instructing SageMaker to distribute or shard the mini batches across all of the compute resources. SageMaker, in turn, shards the training data into all the GPUs and, from time to time, the CPUs on all of the instance types specified in the `estimator` object. So, in order to make this concept easier to understand, *Figure 6.3* illustrates an example of just how SageMaker handles the task:

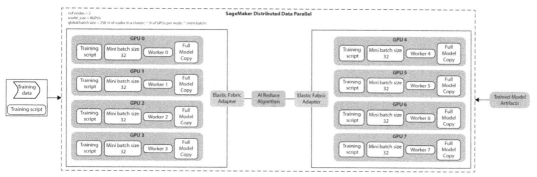

Figure 6.3 – Data parallel training task on SageMaker

As you can see from the example shown in *Figure 6.3* when calling the `fit()` method for the SageMaker `estimator` object, specifying two GPU instances (each with eight GPUs), SageMaker creates a copy of the model training routine, or **training script**, on both of the instances in the ephemeral cluster. Accordingly, each training script is further copied onto each GPU on each of the instances. Once every GPU has a copy of the training script, each GPU, in turn, executes the training script on its individually sharded mini batch of the training data.

The GPU worker then trains the model copy to produce a set of optimal parameters, which then shares with the other GPU workers, both within the same instance, as well as the other GPUs in the second instance, using the **AllReduce optimization algorithm**. This process is repeated in parallel until all of the specified epochs are completed, after which the ephemeral SageMaker cluster is dismantled, and the `fit()` operation is reported as being successful. The result is a single optimized model, stored on S3, and a reduction of the overall model training time, by a factor of 16, which is the total number of GPUs allocated to the task.

So, as you can see, we have effectively reduced the overall training time by implementing a data parallel placement strategy. While this strategy is effective for large training datasets and is a good start at reducing the time it takes to train a model, this strategy, however, doesn't always work when we have large models to train.

> **Note**
>
> Since the model parallel placement strategy is essentially distributing the model's computational graph or model pipeline across multiple nodes, this placement strategy is often referred to as a **pipeline parallel** strategy.

To address the challenge of reducing the overall training time when we have large ML models with millions or even billions of trainable parameters, we can review how to implement a model parallel placement strategy.

Reviewing the SageMaker model data parallel strategy

The data parallel strategy was largely conceived as a method of reducing the overall model training time, where at the time of its induction, training on large quantities of data imposed the biggest challenge. However, with the invention of large-scale **natural language processing** (NLP) models, such as **Generative Pre-trained Transformers** (GPT) from OpenAI (https://openai.com/blog/gpt-3-apps/), training a large ML model will billions of parameters now imposes the biggest challenge. Basically, these models are far too large to fit into the GPU's onboard memory.

Now that we have a rudimentary idea of just how SageMaker implements a data parallel placement strategy, it's relatively easy to translate the concept to a model parallel placement strategy. The key difference is that while a data parallel strategy breaks the large quantity of training data into smaller shards, the model parallel placement strategy performs a similar trick to a large ML model, allowing these smaller pieces of the model to fit into GPU memory. This also means that we don't have to degrade the model's capabilities by having to prune or compress it.

Figure 6.4 highlights just how similar the model parallel execution is to a data parallel execution when SageMaker executes a Training job, using a model parallel placement strategy:

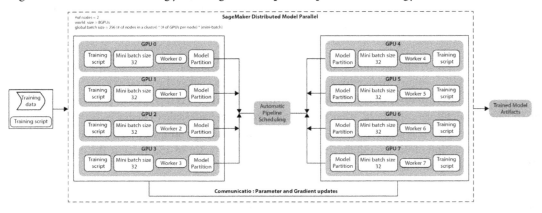

Figure 6.4 – Model parallel training task on SageMaker

You can see from *Figure 6.4* that when calling the `fit()` method for the SageMaker `estimator` object, just like the data parallel example in *Figure 6.3*, SageMaker allocates a copy of the model training script to each GPU on each of the two instances. However, instead of distributing the same copy of the model to each GPU worker, as was the case with the data parallel placement strategy, SageMaker splits the model into smaller pieces, or model partitions, assigning each model partition to a GPU worker.

To coordinate how the training of each model partition, SageMaker implements a **pipeline scheduler**. In the same way that the AllReduce optimization algorithm coordinates parameter optimizations across different GPU workers, the pipeline scheduler ensures that as each batch of data is fed into the model and computation for each partition of the model is correctly coordinated and scheduled between all the GPU workers. This ensures that the matrix calculations for each layer, during both the forward and backward passes over these network partitions, happen in accordance with the overall structure of the model architecture. For example, the scheduler would ensure that the mathematical calculations for layer two are executed before layer three on the forward pass and that layer three's gradient calculation occurs before layer two. Once all the desired epochs have been executed, essentially both the forward and backward passes through the model architecture over the entirety of the training data, SageMaker dismantles the ephemeral cluster and stores the optimized model on S3.

In summation, the data parallel placement strategy was originally conceived to reduce the overall training time of a model by sharing the data and parallelizing the execution across multiple compute resources. The primary motivation behind a model parallel placement strategy is to address large models that don't fit into the compute resource's memory. This then begs the question as to whether it's possible to combine both the data parallel and model parallel placement strategies to reduce the overall training time for both large models, as well as large datasets in a distributed fashion.

Let's review this hybrid methodology next.

Reviewing a hybrid data parallel and model parallel strategy

The fact that both the data parallel and model parallel placement strategies were created to address specific challenges, it is fundamentally impossible to combine both strategies into a unified, hybrid strategy. Essentially, both strategies only solve their specific issues by either sharding the data or sharding the model.

Fortunately, because the entirety of the Training job is orchestrated and managed by SageMaker, the ability to combine both strategies into a hybrid strategy is now possible. For example, if we review *Figure 6.5*, we can visualize just how SageMaker allows us to execute a Training job that implements both the data parallel and model parallel placement strategies independently:

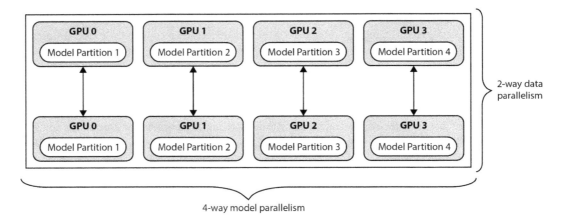

Figure 6.5 – Independent data parallel and model parallel strategies on SageMaker

Figure 6.5 illustrates taking the same number of compute instances, and essentially implementing a two-node data parallel placement strategy, along with a four-way model parallel placement strategy. This translates to creating two copies of the training script and assigning each copy to one of the two compute instances. We then execute the distributed training task using the data parallel placement strategy. While the training task is being executed, we further partition the specific copy of the model architecture across each of the GPUs in the individual compute instance, using the model parallel placement strategy.

So, while each of these placement strategies is unique in its approach, by using SageMaker, we can reap the benefits of both approaches to reduce the overall training time on large datasets, as well as large ML models. In the next section, we will review examples of how to practically implement each of the placement strategies on SageMaker, including an example of this hybrid approach.

Executing a distributed training workload on AWS

Now that we've been introduced to some of the fundamentals of distributed training and what happens behind the scenes when we leverage SageMaker to launch a distributed Training job, let's explore how we can execute such a workload on AWS. Since we've reviewed two placement techniques, namely data parallel and model parallel, we will start by reviewing how to execute distributed data parallel training. After which, we will then review how to execute distributed model parallel training, but also include the hybrid methodology and include an independent data parallel placement strategy alongside the model parallel example.

> **Note**
>
> In this example, we leverage a **Vision Transformer** (**ViT**) model to address an image classification use case. Since the objective of this section is to showcase how to practically implement both the data parallel and model parallel placement strategies, we will not be diving into the particulars of the model itself but rather using it within the context of transfer learning. To learn more about the ViT model, please review the *Transformers for Image Recognition at Scale* paper (`https://arxiv.org/pdf/2010.11929.pdf`).

Let's get started with the data parallel workload.

Executing distributed data parallel training on Amazon SageMaker

There are two crucial elements to executing a distributed Training job using the data parallel placement strategy on SageMaker:

1. Configuring the backend cluster
2. Configuring the model training script

In the next section, we will start by walking through an example of how to configure the backend ephemeral SageMaker cluster.

Configuring the backend cluster

To get started with setting up the SageMaker cluster, we will be leveraging the same SageMaker Studio environment, along with the sample code from the companion GitHub repository, that we introduced in *Chapter 5*, *Data Analysis*.

> **Note**
>
> If you haven't provisioned the SageMaker Studio environment, please refer back to the *Setting up EMR and SageMaker Studio* section in *Chapter 5*, *Data Analysis*.

The following steps will walk you through setting up the example:

1. Log into the AWS account that was used for *Chapter 5*, *Data Analysis*, examples, and open the SageMaker management console (`https://console.aws.amazon.com/sagemaker/home`).

2. With the SageMaker management console open, use the left-hand navigation panel to click on the **SageMaker Domain** link. Under the **Users** section, you will see **Name** of the user, and the **Launch app** drop-down box.

3. Click the **Launch app** drop-down and select the **Studio** option to launch the Studio IDE.

4. Once the Studio environment is open, double-click on the **Applied-Machine-Learning-and-High-Performance-Computing-on-AWS** folder that we cloned in *Chapter 5, Data Analysis*.

5. Now, double-click on the Chapter06 folder to get access to the example Jupyter notebooks.

6. Double-click on the 1_distributed_data_parallel_training.ipynb file to launch the notebook.

Note

The notebook will initialize an **ml.m5.xlarge** compute instance with **4 vCPUs**, and **16 GB** of RAM to run a pre-configured **PyTorch 1.8** kernel. This instance type exceeds the free resource type allowed by the AWS Free Tier (https://aws.amazon.com/free) and will therefore incur AWS usage costs.

---

7. Once the example notebook has been launched and the kernel has been started, click on the **Kernel** menu option, and select the **Restart Kernel and Run All Cells...** option to execute the notebook code cells.

While the notebook is running, let's review the code to understand exactly what's happening. In the first two code cells, we download both the training and validation horses or humans datasets. These datasets have been provided by Laurence Moroney (https://laurencemoroney.com/datasets.html) and contain 500 rendered images of various species of horse, as well as 527 rendered images of humans. We will be using this dataset to generate higher resolution versions, thereby creating much larger image file sizes to simulate having a large training and validation dataset. By increasing the size of the data, we are therefore creating a scenario where training a model on these large image files would, in effect, introduce a delay in the overall time it takes to train an image classification model. Consequently, we are setting up the requirement to leverage a data parallel placement strategy that will, in effect, reduce the overall time taken to train our image classification model.

Note

These datasets are licensed under the Creative Commons 2.0 Attribution 2.0 Unported License.

In the third code cell, as shown in the following code snippet, we programmatically extract both the downloaded train.zip and validation.zip files and save them locally to a data folder:

```
...
import zipfile
with zipfile.ZipFile("train.zip","r") as train_zip_ref:
    train_zip_ref.extractall("data/train")

with zipfile.ZipFile("validation.zip","r") as val_zip_ref:
```

```
        val_zip_ref.extractall("data/validation")
...
```

Now that data has been downloaded and extracted, we should have two folders within the `data` directory called `train` and `validation`. Each of these folders contains images of both horses and humans. However, as already mentioned, these images are pretty small in size. For example, if we examine the `horse01-0.png` file in the `./data/train/horses` folder, you will note that the file is only 151.7 KB in size. Since we only have 500 of these tiny files representing horses, we need to somehow come up with a way to make these files bigger. Therefore, we will use an ML model called **Enhanced Deep Residual Networks for Single Image Super-Resolution** (**EDSR**) to increase the resolution of these files and, in effect, increase the size of the files to simulate a real-world use case where images are in MB, as opposed to KB.

> **Note**
>
> While it's not within the scope of this chapter to detail the EDSR model, we are simply using it to enhance the resolution of the images, thereby making the file size bigger. You can learn more about the pre-trained model from Hugging Face by referencing their model repository (`https://huggingface.co/eugenesiow/edsr-base`).

So, in the next set of code cells, as shown in the following code snippet, we run the pre-trained EDSR model on our image dataset to increase the image resolution and, as a byproduct, increase the image file size:

```
...
from super_image import EdsrModel, ImageLoader
from PIL import Image
import requests
import os
from os import listdir


folder_dir = "data/validation/"
model = EdsrModel.from_pretrained('eugenesiow/edsr-base',
scale=4)
for folder in os.listdir(folder_dir):
    folder_path = f'{folder_dir}{folder}'
    for image_file in os.listdir(folder_path):
        path = f'{folder_path}/{image_file}'
        image = Image.open(path)
        inputs = ImageLoader.load_image(image)
```

```
        preds = model(inputs)
        ImageLoader.save_image(preds, path)

file_size = os.path.getsize(path)
print("File Size is :", file_size/1000000, "MB")
...
```

As you can see from the example output from the code cell, we have increased the file size of each image from approximately 178 KB to just under 2 MB. So, with the datasets ready for training, we can upload them to S3 so that the ephemeral SageMaker cluster can access them. The following code snippet shows how we initialize the SageMaker permissions to S3 and use the `upload()` method from the SageMaker Python SDK's `S3Upload` class to store the data on S3:

```
...
import sagemaker
from sagemaker import get_execution_role
from sagemaker.estimator import Estimator
from sagemaker.s3 import S3Uploader
import boto3

sagemaker_session = sagemaker.Session()
bucket = sagemaker_session.default_bucket()
prefix = 'horse-or-human'
role = get_execution_role()
client = boto3.client('sts')
account = client.get_caller_identity()['Account']
print(f'AWS account:{account}')
session = boto3.session.Session()
region = session.region_name
print(f'AWS region:{region}')
s3_train_data = S3Uploader.upload('data/train',f's3://{bucket}/
{prefix}/data/train')
s3_val_data = S3Uploader.upload('data/validation',f's3://
{bucket}/{prefix}/data/validation')
print('s3 train data path: ', s3_train_data)
print('s3 validation data path: ', s3_val_data)
...
```

Now, we are ready to define the SageMaker estimator. You will recall from the code snippet shown in the *Building ML systems using AWS* section that all we had to do was define a `PyTorch` estimator and provide the basic configuration parameters, such as `instance_count` and `instance_type`, and SageMaker took care of the rest of the heavy lifting to orchestrate the Training job. However, in order to configure a data parallel placement strategy, we need to provide an additional configuration parameter, called `distribution`, to the estimator. As you can see from the following code snippet, we declared the same instance of the estimator, but now we've added the `distribution` parameter to inform SageMaker that we wish to enable a `dataparallel` placement strategy:

```python
...
estimator = PyTorch(entry_point='train.py',
                    source_dir='src',
                    role=role,
                    instance_count=1,
                    instance_type='ml.p3.16xlarge',
                    framework_version='1.8.0',
                    py_version='py3',
                    sagemaker_session=sagemaker_session,
                    hyperparameters={'epochs':10,
                                     'batch_size':32,
                                     'lr':3e-5,
                                     'gamma': 0.7},
                    distribution={"smdistributed":
{"dataparallel": {"enabled": True}}},
                    debugger_hook_config=False,
                    metric_definitions=metric_definitions,
                    )
...
```

Now, all that's left to do is initiate the Training job by calling the `fit()` method of our `estimator` object. The following code snippet shows how to initialize the distributed Training job using the training and validation data that we've already uploaded to S3:

```python
...
from sagemaker.inputs import TrainingInput

train = TrainingInput(s3_train_data,
                      content_type='image/png',
                      input_mode='File')
```

```
val = TrainingInput(s3_val_data,
                    content_type='image/png',
                    input_mode='File')
estimator.fit({'train':train, 'val': val})
...
```

Once the Training job has been initialized, SageMaker will redirect the logs so that we can see what's happening inside the PyTorch training container, and we can match the log output to what we learned about how the distributed data parallel placement strategy functions.

> **Note**
>
> If you receive a **ResourceLimitExceeded** error when calling the estimator.fit() method, you can follow the resolution steps from the *How do I resolve the ResourceLimitExceeded error in Amazon SageMaker?* knowledge article (https://aws.amazon.com/premiumsupport/knowledge-center/resourcelimitexceeded-sagemaker/).

You will recall from the *Reviewing the SageMaker distributed data parallel strategy* section that the training script, as well as the model algorithm, are copied to each GPU within the compute instance. Since each GPU is essentially executing its own copy of the training script to optimize its own unique set of model parameters and then share them with the GPU works using AllReduce, we also need to ensure that the training script itself is configured in such a way that it gets executed as a part of a larger distributed training process.

Basically, what this means is that when we specify the distribution parameter for the SageMaker estimator object, we are instructing SageMaker to configure the appropriate backend resources for a distributed Training job. But we also need to configure the training script to properly use this distributed backend cluster. So, to extend the training script's ability to correctly leverage the backend cluster's distributed capabilities, AWS provides the **Distributed Data Parallel Library**, called smdistributed, for the specified deep learning framework, which is PyTorch in this case.

> **Note**
>
> AWS also provides the smdistributed library for the TensorFlow 2.x deep learning framework. For more information on how to leverage a distributed data parallel placement strategy for a TensorFlow training script, you can review the TensorFlow Guide (https://sagemaker.readthedocs.io/en/stable/api/training/sdp_versions/latest/smd_data_parallel_tensorflow.html).

In the next section, we will review how to configure the model training script using the smdistributed Python library.

Configuring the model training script

There are five basic specific steps for incorporating the smdistributed library into a PyTorch training script. To review these steps, we can open the ./src/train.py file within the Studio IDE and walk through the important code as follows:

1. The first step is to import the smdistributed libraries for the PyTorch framework. As you can see from the following code snippet, by importing and then initializing these modules, we are essentially wrapping PyTorch's ability to execute parallel training methods into the data parallel placement strategy:

   ```
   . . .
   from smdistributed.dataparallel.torch.parallel.
   distributed import DistributedDataParallel as DDP
   import smdistributed.dataparallel.torch.distributed as
   dist
   dist.init_process_group()
   . . .
   ```

2. The next step is to integrate the data into PyTorch's data loading mechanism so that PyTorch can iterate through the chunks of data assigned to the GPU worker. In the following code snippet, we specify num_replicas as the number of GPU workers participating in this distributed training exercise. We also supply the GPU worker's local rank or its membership ranking within the current exercise, by specifying the rank parameter:

   ```
   . . .
   train_sampler = torch.utils.data.distributed.
   DistributedSampler(
           train_dataset, num_replicas=world_size, rank=rank
   )
   . . .
   ```

3. From the previous code snippet, we used the world_size and rank variables. The world_size variable was used to denote the total number of GPU workers over which the data parallel task is being distributed. So, as you can see from the next code snippet, to get the total amount of GPUs, we call the get_world_size() method from the smdistributed. dataparallel.torch.distributed module. Similarly, we also use the get_rank() method from this library to get the current GPU's membership ranking:

   ```
   . . .
   world_size = dist.get_world_size()
   rank = dist.get_rank()
   . . .
   ```

4. Lastly, we configured the mini batch size for the PyTorch `DataLoader()` method to sample, declared as the `batch_size` variable. This is the global batch size of the training job, divided by the number of GPU workers, represented by the `world_size` variable described in *step 3*:

```
...
args.batch_size //= world_size // 8
args.batch_size = max(args.batch_size, 1)
...
```

So, with these minimal code additions applied to the model training routine, we have effectively implemented an example of a data parallel placement strategy. Next, let's look at how to use the same example but apply a model parallel placement strategy.

Executing distributed model parallel training on Amazon SageMaker

Since we are using the same image classification model, shown in the previous example, to illustrate an example of model parallel training, you can follow the same steps to open and execute the notebook. However, instead of opening the `1_distributed_data_parallel_training.ipynb` file, in this example, we are going to open the `2_distributed_model_parallel_training.ipynb` file and run all of the code cells.

So, just as with the data parallel placement strategy, there are two crucial components to successfully implementing the model parallel placement strategy on SageMaker, namely configuring the backend cluster and configuring the model training script. Let's start by exploring all the changes that need to be made to the `estimator` configuration.

Configuring the backend cluster

When reviewing the estimator configuration, note that the options provided to the `distribution` parameter have changed. As you can see from the following code snippet, we now specify a `modelparallel` option instead of enabling the `dataparallel` setting for `smdistributed`:

```
...
                    distribution={
                        "smdistributed": {"modelparallel": smp_
options},
                        "mpi": mpi_options
                    },
...
```

Additionally, as shown in the following code snippet, we declare a variable called `smp_options`, whereby we specify a dictionary of the configuration options specific to the `modelparallel` strategy:

```
...
smp_options = {
    "enabled":True,
    "parameters": {
        "partitions": 1,
        "placement_strategy": "spread",
        "pipeline": "interleaved",
        "optimize": "speed",
        "ddp": True,
    }
}

mpi_options = {
    "enabled" : True,
    "processes_per_host" : 8,
}
...
```

As you can see from the previous code snippet, where the most important configuration options have been highlighted, we set the `placement_strategy` parameter as `spread`. In effect, we are configuring SageMaker to evenly spread the model partitions across all GPU devices within the compute instance. Since we are using a single **ml.p3.16xlarge** instance with eight GPUs, and not multiple compute instances, we are spreading the model partitions evenly within the instance.

Additionally, we are setting the pipeline scheduling mechanism, the `pipeline` parameter, to `interleaved`. This setting improves the overall performance of the backend cluster by prioritizing the backward execution model exaction tasks to free up GPU memory.

Lastly, to enable both a model parallel, as well as a hybrid implementation of the data parallel placement strategies, we set the distributed data parallel, or `ddp` parameter, to `True`. As we saw in the section entitled, *Reviewing a hybrid data parallel and model parallel strategy*, both the data parallel and model parallel strategies can be used at the same time to further reduce the overall time it takes to train the model.

So, since we are using both strategies concurrently for this example, we must also supply a **Message Passing Interface** (**MPI**), declared as the `mpi` parameter, to instruct SageMaker as to how each GPU worker communicates what it's doing with the other GPU workers. For example, in the previous code snippet, after enabling the `mpi_options` setting, we have also set `processes_per_host` to 8. This setting, in effect, configures the ephemeral SageMaker cluster architecture to match *Figure 6.5*, where we set the GPU workers on the single **ml.p3.16xlarge** compute instance to use a four-way model parallel strategy to essentially partition the model across four GPU workers. Additionally, we also configure a two-way data parallel strategy to partition the training data into two shards and execute the model partitions in parallel across the shards. Therefore, two-way x four-way equates to eight processes per single host.

As you can see, adding these minimal configuration changes implements a data parallel and model parallel capable SageMaker cluster. Yet, just as with the previous example, there are also changes that need to be made to the training script. Let's review these next.

Configuring the model training script

Since implementing a model parallel placement strategy is more intricate than a data parallel strategy, there are a few extra requirements that need to be added to the training script. Let's now open the `./src/train_smp.py` file to review the most important requirements. As you might immediately notice, there are 11 specific script changes required to execute a model parallel placement strategy for a PyTorch model:

1. Once again, and as you can see from the following code snippet, the first step is to import the `modelparallel` modules from the `smdistributed` library and initialize these modules as a wrapper for PyTorch:

    ```
    ...
    import smdistributed.modelparallel
    import smdistributed.modelparallel.torch as smp
    smp.init()
    ...
    ```

2. Once the module has been initialized, we then extend our image classification model, defined using the `model` variable, and wrap it into the `DistributedModel()` class, as shown in the following code block. This signals that our model is now being distributed:

    ```
    ...
    model = smp.DistributedModel(model)
    ...
    ```

3. Since the model is now being distributed, we also need to distribute the optimizer. So, as you can see from the following code snippet, we optimize the model parameters using PyTorch's implementation of the `Adam()` algorithm and subsequently distribute the optimization task across GPUs by wrapping it into the `DistributedOptimizer()` class:

    ```
    ...
    optimizer = smp.DistributedOptimizer(
    ```

```
          optim.Adam(model.parameters(), lr=args.lr))
    . . .
```

4. Alongside distributing the model itself, as well as the optimizer, we also need to define exactly how the forward and backward pass through the model's computational graph, or the model pipeline, are executed. Accordingly, we extend the computation results from both the forward and backward passes of the model by wrapping them within a `step()` decorator function. The following code snippet shows the `step()` decorator that extends `train_step()` for the forward pass, and `test_step()` for the backward pass:

```
    . . .
    @smp.step
    def train_step(model, data, label):
        output = model(data)
        loss = F.nll_loss(F.log_softmax(output), label,
                          reduction="mean")
        # replace loss.backward() with model.backward in the
    train_step function.
        model.backward(loss)
        return output, loss


    @smp.step
    def test_step(model, data, label):
        val_output = model(data)
        val_loss = F.nll_loss(F.log_softmax(val_output),
                          label, reduction="mean")
        return val_loss
    . . .
```

5. Lastly, once the model has been trained using the model parallel strategy, and as you can see from the following code snippet, we only save the final model on the highest-ranking GPU worker of the cluster:

```
    . . .
    if smp.rank() == 0:
        model_save = model.module if hasattr(model,
    "module") else model
        save_model(model_save, args.model_dir)
    . . .
```

While these are only a few of the most important parameters for configuring the training script using the `smdistributed.modelprallel` module, you can see that with a minimal amount of code, we can fully provision our training script to use an automatically configured SageMaker ephemeral cluster for both a data parallel and model parallel placement strategy, thus reducing the overall training time using this hybrid implementation.

Summary

In this chapter, we drew your attention to two potential challenges that ML practitioners may face when training ML models: firstly, the challenge of reducing the overall model training time, especially when there is a large amount of training data; and secondly, the challenge of reducing the overall model training time when there are large models with millions and billions of trainable parameters.

We reviewed three specific strategies that can be used to address these challenges, namely the data parallel placement strategy, which distributes a large amount of training data across multiple worker resources to execute the model training process in parallel. Additionally, we also reviewed the model parallel placement strategy, which distributes a very large ML model across multiple GPU resources to offset trying to squeeze these large models into the available memory resources. Lastly, we also explored how both these strategies can be combined, using a hybrid methodology, to further reap the benefits that both offer.

Furthermore, we also reviewed how Amazon SageMaker can be used to solve these challenges, specifically focusing on how SageMaker takes care of the heavy lifting of building a distributed training compute and storage infrastructure specifically configured to handle any of these three placement strategies. SageMaker not only provisions the ephemeral compute resources but also provides Python libraries that can be integrated into the model training script to fully make use of the cluster.

Now that we've seen how to carry out ML model training using distributed training, in the next chapter, we will review how to deploy the trained ML models at scale.

7

Deploying Machine Learning Models at Scale

In previous chapters, we learned about how to store data, carry out data processing, and perform model training for machine learning applications. After training a machine learning model and validating it using a test dataset, the next task is generally to perform inference on new and unseen data. It is important for any machine learning application that the trained model should generalize well for unseen data to avoid overfitting. In addition, for real-time applications, the model should be able to carry out inference with minimal latency while accessing all the relevant data (both new and stored) needed for the model to do inference. Also, the compute resources associated with the model should be able to scale up or down depending on the number of inference requests, in order to optimize cost while not sacrificing performance and inference requirements for real-time machine learning applications.

For use cases that do not require real-time inference, the trained model should be able to carry out inference on very large datasets with thousands of variables in a reasonable amount of time as well. In addition, in several scenarios, we may not want to go through the effort of managing servers and software packages needed for inference, and instead, focus our effort on developing and improving our machine learning models.

Keeping all these aforementioned factors in mind, AWS provides multiple options for deploying machine learning models to carry out inference on new and unseen data. These options consist of real-time inference, batch inference, and asynchronous inference. In this chapter, we are going to discuss the managed deployment options of machine learning models using Amazon SageMaker, along with various features such as high availability of models, auto-scaling, and blue/green deployments.

We will cover the following topics in this chapter:

- Managed deployment on AWS
- Choosing the right deployment option
- Batch inference
- Real-time inference
- Asynchronous inference

- High availability of model endpoints
- Blue/green deployments

Managed deployment on AWS

Data scientists and machine learning practitioners working on developing machine learning models to solve business problems are often very focused on model development. Problem formulation and developing an elegant solution, choosing the right algorithm, and training the model so that it provides reliable and accurate results are the main components of the machine learning problem solving cycle that we want our data scientists and data engineers to focus on.

However, once we have a good model, we want to run real-time or batch inference on new data. Deploying the model and then managing it are tasks that often require dedicated engineers and computation resources. This is because, we first need to make sure that we have all the right packages and libraries for the model to work correctly. Then, we also need to decide on the type and amount of compute resources needed for the model to run. In real-time applications, we often end up designing for peak performance requirements, just like provisioning servers for IT projects.

After the model is deployed and is running, we also need to make sure that everything stays working in the manner that we expect it to. Furthermore, in real world scenarios, data scientists often have to manually carry out analysis periodically to detect model or data drift. In the event that either of these drifts are detected, the data scientists go through the entire cycle of exploratory analysis, feature engineering, model development, model training, hyperparameter optimization, model evaluation, and model deployment again. All these tasks consume a lot of effort and resources and due to this reason, many organizations have moved to automating these processes using **machine learning operations** (**MLOps**) workflows and managed model deployment options that scale well with varying workloads.

Amazon SageMaker offers multiple fully managed model deployment options. In this section, we give an overview of these managed deployment options along with their benefits and then discuss a few of these deployment options in detail in the following sections.

Amazon SageMaker managed model deployment options

Amazon SageMaker offers the following managed deployment model options:

- **Batch Transform**: SageMaker Batch Transform is used to carry out inference on large datasets. There is no persistent endpoint in this case. This method is commonly used to carry out inference in a non-real-time machine learning use case requiring offline inference on larger datasets.

- **Real-time endpoint**: A SageMaker real-time endpoint is for use cases where a persistent machine learning model endpoint is needed, which carries out inference on a few data samples in real time.

- **Asynchronous inference**: Amazon SageMaker Asynchronous Inference deploys an asynchronous endpoint for carrying out inference on large payloads (up to 1 GB) with large processing times and low latency.

- **Serverless Inference**: In all the previous methods, the user is required to select the compute instance types for inference. Amazon SageMaker Serverless Inference automatically chooses the server type and scales up and down based on the load on the endpoint. It is often useful for applications that have unpredictable traffic patterns.

Let's explore the variety of available compute resources next.

The variety of compute resources available

To carry out inference, there are a wide variety of computation instances available. At the time of writing, approximately 70+ instances are available for carrying out machine learning inference. These instances have varying levels of computation power and memory available to serve different use cases. There is also the option of using **graphical processing units** (**GPUs**) for inference. In addition, SageMaker also supports Infl instances for high-performance and low-cost inference. These options make SageMaker model deployment and inference highly versatile and suitable for a variety of machine learning use cases.

Cost-effective model deployment

Amazon SageMaker has various options for optimizing model deployment cost. There are multi-model endpoints, where multiple models can share a container. This helps with reducing hosting costs since endpoint utilization is increased due to multiple models sharing the same endpoint. In addition, it also enables time sharing of memory resources across different models. SageMaker also has the option of building and deploying multi-container endpoints. Furthermore, we can attach scaling policies to our endpoints to allocate more compute resources when traffic increases, and shut down instances when traffic decreases in order to save on costs.

Another cost-effective option for model deployment is SageMaker Serverless Inference. Serverless Inference utilizes AWS Lambda to scale up and down compute resources as traffic increases or decreases. It is especially useful for scenarios with unpredictable traffic patterns.

Blue/green deployments

SageMaker automatically uses blue/green deployment whenever we update a SageMaker model endpoint. In blue/green deployment, SageMaker uses a new fleet of instances to deploy the updated endpoints and then shifts the traffic to the updated endpoint from the old fleet to the new one. Amazon SageMaker offers the following traffic-shifting strategies for blue/green deployments:

- All at once traffic shifting
- Canary traffic shifting
- Linear traffic shifting

We will discuss these traffic patterns in more detail later in this chapter.

Inference recommender

With Amazon SageMaker Inference Recommender, we can automatically get recommendations on the type of compute instance to use for deploying our model endpoint. It gives us instance recommendations by load testing various instances and outputs inference costs, along with throughput and latency, for the tested instance types. This helps us decide on the type of instance to use for deploying our model endpoint.

MLOps integration

Using Amazon SageMaker, we can easily build machine learning workflows that integrate with **continuous integration and continuous delivery (CI/CD)** pipelines. These workflows can be used to automate the entire machine learning life cycle, including data labeling, data processing and feature engineering, model training and registry, post-processing, endpoint deployment, and model monitoring for data and model drift monitoring. For model deployment, these workflows can be used to automate the process of doing batch inference, as well as pushing model endpoints from development to staging to production environments.

Model registry

Amazon SageMaker provides the capability to register and catalog machine learning models with the SageMaker model registry. Using the model registry, we can include different versions of a trained model in a model package group. This way, whenever we train and register a model, it is added as a new version to the model package group. In addition, using the model registry, we can also associate metadata and training metrics to a machine learning model, approve or reject a model, and if approved, move the models to production. These features of the model registry facilitate the building of CI/CD pipelines needed for automating machine learning workflows.

Elastic inference

For machine learning use cases that require very high throughput and low latency, we often end up using GPU machines, thereby increasing the cost of inference significantly. Using Amazon SageMaker, we can add elastic inference to our endpoints. Elastic inference provides inference acceleration to our endpoint, by attaching just the right amount of GPU-powered inference acceleration to any SageMaker instance type. This helps significantly with latency and throughput, while also achieving it at a much lower cost compared to using GPU instances for inference.

Deployment on edge devices

Many machine learning use cases require models to run on edge devices such as mobile devices, cameras, and specialized devices. These devices often have low compute resources, memory, and storage. Furthermore, deploying, managing, and monitoring machine learning models on a fleet of devices is a difficult task because of the variability in device hardware and operating systems. With Amazon SageMaker Edge Manager, machine learning models can be deployed, monitored, and managed on a fleet of devices with different hardware and software configurations. SageMaker Edge Manager uses SageMaker Neo to compile machine learning models and packages these compiled models to be deployed on edge devices. In addition, we can also sample the data used by the model on the edge devices and send them to the cloud to carry out analysis to determine quality issues such as data and model drift.

Now, let's discuss the various model deployment options on AWS in the following section.

Choosing the right deployment option

As mentioned in the previous section, AWS has multiple model deployment and inference options. It can get confusing and overwhelming sometimes to decide on the right option for model deployment. The decision to select the right model deployment option really depends on the use case parameters and requirements. A few important factors to consider while deciding on deployment options are listed as follows:

- Do we have an application that needs a real-time, persistent endpoint to carry out on-demand inference on new data in real time and very quickly with low latency and high availability?

- Can our application wait for a minute or two for the compute resources to come online before getting the inference results?

- Do we have a use case where we do not need results in near real time? Can we do inference on a batch of data once a day/week or on an as-needed basis?

- Do we have an unpredictable and non-uniform traffic pattern requiring inference? Do we need to scale up and down our compute resources based on the traffic?

- How big is the data (number of data points/rows) that we are trying to do inference on?

- Do we need dedicated resources all the time to carry out inference or can we follow a serverless approach?

- Can we pack in multiple models in a single endpoint to save on cost?

- Do we need to have an inference pipeline consisting of multiple models, pre-processing steps, and post-processing steps?

In the following subsections, we will discuss when to pick the different types of model deployment and inference options provided by Amazon SageMaker, while addressing the previously mentioned questions. We will also provide examples of typical example use cases for each of the model deployment options.

Using batch inference

Amazon SageMaker Batch Transform is used when there is no need for a persistent, real-time endpoint and inference can be done on large batches of data. The following examples illustrate the use of SageMaker Batch Transform:

- **Predictive maintenance**: In a manufacturing plant, sensor and machine data for various components could be collected the entire day. For such a use case, there is no need for real-time or asynchronous endpoints. At night, machine learning models can be used for predicting whether a component is about to fail, or whether a part of the machinery needs maintenance. These models would run on large batches of data and carry out inference using SageMaker Batch Transform. The results from these models could then be used to make and execute a maintenance schedule.

- **Home prices prediction**: Real estate companies collect data for a few days (and sometimes weeks) before coming out with new home prices and market direction predictions. These models do not need real-time or asynchronous endpoints as they need to be run only after a few days or weeks and on large amounts of data. For such use cases, SageMaker Batch Transform is the ideal option for inference. SageMaker Batch Transform jobs could run on a fixed interval in a machine learning pipeline on new and historical data, carrying out inference to predict home price adjustments and market direction by localities. These results can, in turn, then be used to determine if the machine learning models need to be retrained.

We will cover batch, real-time, and asynchronous inference options in Amazon SageMaker in detail in the later sections of this chapter.

Using real-time endpoints

Amazon SageMaker real-time endpoints should be the choice for model deployment when there is a need for a real-time persistent model endpoint, doing predictions with low latency as new data arrives. Real-time endpoints are fully managed by Amazon SageMaker and can be deployed as multi-model and multi-container endpoints. The following are some example use cases for real-time endpoints:

- **Fraudulent transaction**: A customer uses a credit card to purchase an item online or physically in a retail store. This financial transaction can be carried out by the actual owner of the credit card or it can be a stolen credit card as well. The financial institution needs to make the decision in real time whether to approve the transaction or not. In such a scenario, a machine learning model can be deployed as a real-time endpoint. This model could use customer's demographic data and history of purchases from historical data tables, while also using some data from the current transaction, such as IP address and web browser parameters (if it is an online transaction), or store location and image and/or video from a camera in real time (if it is a physical transaction), to classify whether the transaction is fraudulent or not. This decision can then be used by the financial institution to either approve or reject the transaction, or contact the customer for notification, manual authentication, and approval.

- **Real-time sentiment analysis**: A customer is having a chat or phone conversation with a customer care agent. A machine learning model is analyzing the chat or transcribed text from voice conversation in real time to decide on the sentiment that the customer is showing. Based on the sentiment, if the customer is unhappy, the agent can offer various promotions or escalate the case to a supervisor before things get out of hand. This machine learning model should be deployed using a real-time endpoint so that the sentiment can be correctly determined without any lag or delay.

- **Quality assurance**: In a manufacturing plant, products are being assembled on an assembly line and there needs to be strict quality control to remove defective products as soon as possible. This is again an application where real-time inference from a machine learning model classifying the objects as defective or normal using live images or video feed will be useful.

Similar to real-time endpoints, we also have the option of using asynchronous endpoints, which we will learn about in the following section.

Using asynchronous inference

Amazon SageMaker Asynchronous Inference endpoints are very similar to real-time endpoints. Asynchronous endpoints can queue inference requests and are the deployment option of choice when near real-time latency is needed, while also processing large workloads. The following example illustrates a potential asynchronous endpoints use case.

Train track inspection: Several trains are running on their everyday routes with attached cameras that take images and videos of train tracks and switches for defect detection. These trains do not have a high bandwidth available to transmit this data in real time for inference. When these trains dock at a station, a large number of images and videos could be sent to a SageMaker Asynchronous Inference endpoint for inference to find out whether everything is normal, or if there are any defects present anywhere on the track or switches. The compute instance associated with the asynchronous endpoint will start as soon as it receives data from any of the trains, carrying out inference on the data, then shutting down once all the data has been processed. This will help with the reduction in costs compared to a real-time endpoint.

Batch inference

For carrying out batch inference on datasets, we can use SageMaker Batch Transform. It should be used for inference when there is no need for a real-time persistent deployed machine learning model. Batch Transform is also useful when the dataset for inference is large or if we need to carry out heavy preprocessing on the dataset. For example, removing bias or noise from the data, converting speech data to text, and filtering and normalization of images and video data.

We can pass input data to SageMaker Batch Transform in either one file or using multiple files. For tabular data in one file, each row in the file is interpreted as one data record. If we have selected more than one instance for carrying out the batch transform job, SageMaker distributes the input files to different instances for batch transform jobs. Individual data files can also be split into multiple mini-batches and batch transform on can be carried out these mini-batches in parallel on separate instances. *Figure 7.1* shows a simplified typical architecture example of SageMaker Batch Transform:

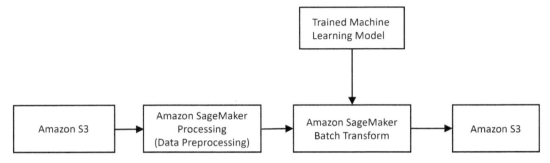

Figure 7.1 – Example architecture for Amazon SageMaker Batch Transform

As shown in the figure, data is read from **Amazon S3**. Preprocessing and feature engineering is carried out on this data using **Amazon SageMaker Processing** in order to transform the data in the right format that the machine learning model expects. A trained machine learning model is then used by **Amazon SageMaker Batch Transform** to carry out batch inference. The results are then written back to **Amazon S3**. The machine learning model used by Batch Transform could either be trained using Amazon SageMaker or can be a model trained outside of Amazon SageMaker. The two main steps in Batch Transform are as follows:

1. Creating a transformer object.
2. Creating a batch transform job for carrying out inference.

These steps are described in the following subsections.

Creating a transformer object

We first need to create an object of the `Transformer` class in order to run a SageMaker batch transform job. While creating the `Transformer` object, we can specify the following parameters:

- `model_name`: This is the name of the machine learning model that we are going to use for inference. This can also be a built-in SageMaker model, for which we can directly use the `transformer` method of the built-in estimator.
- `instance_count`: The number of EC2 instances that we are going to use to run our batch transform job.

- `instance_type`: The type of EC2 instances that we can use. A large variety of instances are available to be used for batch transform jobs. The choice of instance depends on our use case's compute and memory requirements, as well as the data type and size.

In addition, we can also specify several other parameters such as batch strategy, and output path. The complete list of parameters can be found on the documentation page in the *References* section. In the following example, we used SageMaker's built-in XGBoost container. For running batch transformers for your own containers/models or frameworks, such as PyTorch and TensorFlow, the container images may vary.

Figures 7.2 – 7.8 show an example of SageMaker's XGBoost model being fit on our training data and then a transformer object being created for this training model. We specified that the transform job should run on one instance of type `ml.m4.xlarge` and should expect `text/csv` data:

1. As shown in *Figure 7.2*, we can specify the various packages and SageMaker parameters needed to run the code in the example.

```
[1]: #Importing SageMaker Python SDK and other libraries needed to run the code in this example
     import sagemaker
     import re
     import boto3
     from sagemaker import get_execution_role
     import numpy as np
     import pandas as pd
     import matplotlib.pyplot as plt
     from IPython.display import Image
     from IPython.display import display
     from time import gmtime, strftime
     import sys
     import json
     import os
     import zipfile
     bucket=sagemaker.Session().default_bucket()
     prefix = 'hpc-book/model-deployment-demo'

     role = get_execution_role()

[2]: #Downloading the abalone dataset for UC Irvine datasets repositories
     !wget https://archive.ics.uci.edu/ml/machine-learning-databases/abalone/abalone.data
```

Figure 7.2 – Setting up the packages and bucket in SageMaker and downloading
the dataset to be used for model training and inference

2. As shown in *Figure 7.3*, we can read the data and carry out one hot encoding on the categorical variables.

```
[3]: #Reading the data, adding column names and carrying out one hot encoding
     #on categorical data
     data = pd.read_csv('abalone.data', header = None)
     data.columns = ['Sex', 'Length', 'Diameter', 'Height', 'WholeWeight',
                     'ShuckedWeight', 'VisceraWeight', 'ShellWeight','Rings']
     model_data = pd.get_dummies(data)
     model_data
```

	Length	Diameter	Height	WholeWeight	ShuckedWeight	VisceraWeight	ShellWeight	Rings	Sex_F	Sex_I	Sex_M
0	0.455	0.365	0.095	0.5140	0.2245	0.1010	0.1500	15	0	0	1
1	0.350	0.265	0.090	0.2255	0.0995	0.0485	0.0700	7	0	0	1
2	0.530	0.420	0.135	0.6770	0.2565	0.1415	0.2100	9	1	0	0
3	0.440	0.365	0.125	0.5160	0.2155	0.1140	0.1550	10	0	0	1
4	0.330	0.255	0.080	0.2050	0.0895	0.0395	0.0550	7	0	1	0
...	...	...	...	...	...	...	...	...	...	...	...

Figure 7.3 – Doing one hot encoding on categorical variables and displaying the results

3. *Figure 7.4* shows the data being split into training, validation, and testing partitions to be used during the model training process.

```
[4]: #Splitting the data in train, validation and test sets
     train_data, validation_data, test_data = np.split(
         model_data.sample(frac=1, random_state=2513),
         [int(0.7 * len(model_data)), int(0.9 * len(model_data))])
```

Figure 7.4 – Splitting the data into train, validation, and test sets for model training and testing

4. As shown in *Figure 7.5*, we can rearrange the columns in the data table in the order that the machine learning model (XGBoost) expects it to be. Furthermore, we can also upload the training and validation data files to an S3 bucket for the model training step.

```
[5]: #Moving the target variable to the beginning of the data frame since xgboost expects
     #the first column to be target cariable
     pd.concat([train_data['Rings'], train_data.drop(['Rings'], axis=1)],
             axis=1).to_csv('train.csv', index=False, header=False)
     pd.concat([validation_data['Rings'], validation_data.drop(['Rings'], axis=1)],
             axis=1).to_csv('validation.csv', index=False, header=False)

     #Uploading the data to Amazon S3 bucket for saving and model training
     boto3.Session().resource('s3').Bucket(bucket).Object(
         os.path.join(prefix, 'train/train.csv')).upload_file('train.csv')
     boto3.Session().resource('s3').Bucket(bucket).Object(
         os.path.join(prefix, 'validation/validation.csv')).upload_file('validation.csv')
```

Figure 7.5 – Reorganizing the data and uploading to S3 bucket for training

5. *Figure 7.6* shows that we are using SageMaker's XGBoost container for training our model. It also specifies the data channels for training and validating the model.

```
[6]:  #Retrieving SageMaker's XGBoost container for model training
      container = sagemaker.image_uris.retrieve(region=boto3.Session().
                          region_name, framework='xgboost', version='latest')
```

```
[7]:  s3_input_train = sagemaker.inputs.TrainingInput(s3_data=
          's3://{}/{}/train'.format(bucket, prefix), content_type='csv')
      s3_input_validation = sagemaker.inputs.TrainingInput(s3_data=
          's3://{}/{}/validation/'.format(bucket, prefix), content_type='csv')
```

Figure 7.6 – Specifying the container for model training along with
training and validation data path in S3

As shown in *Figure 7.7*, we need to define the estimator, the instance type, and instance count, and set various hyperparameters needed by the XGBoost model. We will also start the training job by calling the `fit` method.

```
[8]:  #Specifying the instance count and type for model training.
      #Also specifying the output #path for model artifacts.
      sess = sagemaker.Session()

      xgb_model = sagemaker.estimator.Estimator(container,
                                    role,
                                    instance_count=1,
                                    instance_type='ml.m4.xlarge',
                                    output_path='s3://{}/{}/output'
                                          .format(bucket, prefix),
                                    sagemaker_session=sess)
      #Setting XGBoost hyperparameters
      xgb_model.set_hyperparameters(max_depth=6,
                              eta=0.3,
                              gamma=4,
                              min_child_weight=7,
                              subsample=0.9,
                              silent=0,
                              objective='reg:linear',
                              num_round=80)

      #Starting thge model training job
      xgb_model.fit({'train': s3_input_train, 'validation': s3_input_validation})
```

Figure 7.7 – Defining the SageMaker estimator for training and setting up the hyperparameters

Next, let's create a batch transform job.

Creating a batch transform job for carrying out inference

After creating the transformer object, we need to create a batch transform job. *Figure 7.8* shows an example of starting a batch transform job using the `transform` method call of the batch transformer object. In this transform call, we will specify the location of data in Amazon S3 on which we want to carry out batch inference. In addition, we will also specify the content type of the data (`text/csv`, in this case), and how the records are split in the data file (each line containing one record, in this case).

```
[9]:  #Defining a transformer for carrying out batch tarnsform on the
      #XGBoost model

      batch_transformer = xgb_model.transformer(1, "ml.m5.xlarge",
                                                accept="text/csv")

      # Preparing the data for the batch transform job
      test_data_for_inference = test_data.drop(columns=["Rings"], axis=1)
      test_data_for_inference.to_csv("test_data_for_inference.csv",
                                     index=False, header=False)
      inference_test_data_s3 = sess.upload_data(
          "test_data_for_inference.csv", bucket=bucket, key_prefix=prefix)

      #Starting the batch transform job
      batch_transformer.transform(inference_test_data_s3,
                                  split_type="Line", content_type="text/csv")
```

Figure 7.8 – Creating a transformer object, preparing data for batch
inference, and starting the batch transform job

Figure 7.9 shows an example of reading the results from S3 and then plotting the results (actual versus predictions):

```
[10]:  #Defining function to read data from S3 bucket
       import json
       import io
       from urllib.parse import urlparse


       def read_data_from_s3(s3uri, file_name):
           s3url = urlparse(s3uri)
           bucket_name = s3url.netloc
           prefix = s3url.path[1:]
           s3 = boto3.resource("s3")
           s3Obj = s3.Object(bucket_name, "{}/{}".format(prefix, file_name))
           return s3Obj.get()["Body"].read().decode("utf-8")

[11]:  output = read_data_from_s3(batch_transformer.output_path,
                                  "test_data_for_inference.csv.out")
       output_df = pd.read_csv(io.StringIO(output), sep=",", header=None)

[25]:  #Plotting the actual target variable along with the prediction
       #values from model inference
       plt.plot(test_data['Rings'], np.round(output_df), '+', color = 'red')
       plt.xlabel('Actual Values')
       plt.ylabel('Predictions');
```

Figure 7.9 – Creating a helper function for reading the results of batch transform, and plotting the results (actual versus predictions)

Figure 7.10 shows the resulting plot:

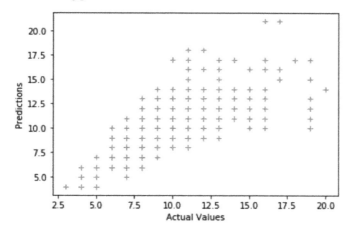

Figure 7.10 – Plot of actual versus prediction Rings values

The code and steps discussed in this section outline the process of training a machine learning model and then carrying out batch inference on the data using the model.

Optimizing a batch transform job

SageMaker Batch Transform also gives us the option of optimizing the transform job using a few hyperparameters, as described here:

- `max_concurrent_transforms`: The maximum number of HTTP requests that can be made to each batch transform container at any given time. To get the best performance, this value should be equal to the number of compute workers that we are using to run our batch transform job.

- `max_payload`: This value specifies the maximum size (in MB) of the payload in a single HTTP request sent to the batch transform for inference.

- `strategy`: This specifies the strategy whether we want to have just one record or multiple records in a batch.

SageMaker Batch Transform is a very useful option to carry out inference on large datasets for use cases that do not require real-time latency and high throughput. The execution of SageMaker Batch Transform can be carried out by using an MLOps workflow, which can be triggered whenever there is new data on which inference needs to be carried out. Automated reports can then be generated using the batch transform job results.

Next, we will learn about deploying a real-time endpoint for making predictions on data using Amazon SageMaker's real-time inference option.

Real-time inference

As discussed earlier in this chapter, the need for real-time inference arises when we need results with very low latency. Several day-to-day use cases are examples of using real-time inference from machine learning models, such as face detection, fraud detection, defect and anomaly detection, and sentiment analysis in live chats. Real-time inference in Amazon SageMaker can be carried out by deploying our model to the SageMaker hosting services as a real-time endpoint. *Figure 7.11* shows a typical SageMaker machine learning workflow of using a real-time endpoint.

Figure 7.11 – Example architecture of a SageMaker real-time endpoint

In this figure, we first read our data from an Amazon S3 bucket. Data preprocessing and feature engineering are carried out on this data using Amazon SageMaker Processing. A machine learning model is then trained on the processed data, followed by results evaluation and post-processing (if any). After that, the model is deployed as a real-time endpoint for carrying out inference on new data in real time with low latency. Also shown in the figure is SageMaker Model Monitor, which is attached to the endpoint in order to detect data and concept drift on new data that is sent to the endpoint for inference. SageMaker also provides the option of registering our machine learning model with the SageMaker Model Registry, for various purposes, such as cataloging, versioning, and automating deployment.

Hosting a machine learning model as a real-time endpoint

Amazon SageMaker provides us with many options to host a model or multiple models as real-time endpoints. We can either use SageMaker Python SDK, the AWS SDK for Python (Boto3), the SageMaker console, or the AWS **command-line interface** (**CLI**) to host our models. Furthermore, these endpoints can also host a single model, multiple models in one container in a single endpoint, and multiple models using different containers in a single endpoint. In addition, a single endpoint can also host models containing preprocessing logic as a serial inference pipeline. We will go through these multiple options in the following subsections.

Single model

As mentioned in the preceding section, we can host a model endpoint using multiple options. Here, we will show you how to host an endpoint containing a single model using the Amazon SageMaker SDK. There are two steps involved in creating a single model endpoint using the Amazon SageMaker SDK, as described here:

1. **Creating a model object**: We need a model object to deploy as an endpoint. The first step to create a real-time endpoint is to use the Model class to create a model object that can be deployed as a HTTPS endpoint. We can also use the model trained in SageMaker. For example, the XGBoost model trained in the example shown in *Figure 7.7*.

2. **Creating the endpoint**: The next step is to use the model object's deploy() method to create an HTTPS endpoint. The deploy method requires the instance type as well as an initial instance count to deploy the model with. This is shown in *Figure 7.12*:

```
[13]:   #Deploying the XGBoost model as a real-time endpoint
        xgb_predictor = xgb_model.deploy(initial_instance_count=1,
                                 instance_type='ml.m4.xlarge')
```

Figure 7.12 – Calling the deploy method of the XGBoost estimator we trained
earlier to deploy our model on a single instance of type ml.m4.xlarge

Figure 7.13 show a custom inference function to serialize the test data, and send it to the real-time endpoint for inference:

```
[14]:   xgb_predictor.serializer = sagemaker.serializers.CSVSerializer()

[15]:   #Defining an inference helper function for carrying out inference
        #on the data sent to the endpoint

        def inference(data, predictor, rows=500 ):
            split_array = np.array_split(data,
                                int(data.shape[0] / float(rows) + 1))
            predictions = ''
            for array in split_array:
                predictions = ','.join([predictions,
                                    predictor.predict(array).decode('utf-8')])

            return np.fromstring(predictions[1:], sep=',')

        #Calling the inference helper function ton carry out inference
        #on the test data
        predictions = inference(test_data.drop(['Rings'], axis=1)
                            .to_numpy(), xgb_predictor)
        predictions = np.round(predictions).astype(int)

[16]:   #Forming data frame with actual values and predictions and
        #displaying the head of the results

        df_output_realtime = pd.DataFrame({'Actual': test_data['Rings'],
                                    'Predictions': predictions})
        df_output_realtime.head()
```

Figure 7.13 – Serializing the data to be sent to the real-time endpoint. Also,
creating a helper function to carry out inference using the endpoint

Figure 7.14 displays the inference results for a few records along with the actual values of our target variable—`Rings`:

Actual	Predictions
10	10
6	7
11	13
13	17
11	10

Figure 7.14 – Showing a few prediction results versus actual values (Rings)

The endpoint will continue incurring costs as long as it is not deleted. Therefore, we should use real-time endpoints only when we have a use case in which we need inference results in real time. For use cases where we can do inference in batches, we should use SageMaker Batch Transform.

Multiple models

For hosting multiple models in a single endpoint, we can use SageMaker multi-model endpoints. These endpoints can be used to also host multiple variants of the same model. Multi-model endpoints are a very cost-effective method of saving our inference cost for real-time endpoints since the endpoint utilization is generally more when we use multi-model endpoints. We can use business logic to decide on the model to use for inference.

With multi-model endpoints, memory resources across models are also shared. This is very useful when our models are comparable in size and latency. If there are models that have significantly different latency requirements or transactions per second, then single model endpoints for the models are recommended. We can create multi-model endpoints using either the SageMaker console or the AWS SDK for Python (Boto3). We can follow similar steps as those for the creation of a single-model endpoint to create a multi-model endpoint, with a few differences. The steps for Boto3 are as follows:

1. First, we need a container supporting multi-model endpoints deployment.

2. Then, we need to create a model that uses this container using Boto3 SageMaker client.

3. For multi-model endpoints, we also need to create an endpoint configuration, specifying instance types and initial counts.

4. Finally, we need to create the endpoint using the `create_endpoint()` API call of the Boto3 SageMaker client.

While invoking a multi-model endpoint, we also need to pass a target model parameter to specify the model that we want to use for inference with the data in the request.

Multiple containers

We can also deploy models that use different containers (such as different frameworks) as multi-container endpoints. These containers can be run individually or can also be run in a sequence as an inference pipeline. Multi-container endpoints also help improve endpoint utilization efficiency, hence cutting down on the cost associated with real-time endpoints. Multi-container endpoints can be created using Boto3. The process to create a multi-container endpoint is very similar to creating multi-model and single-model endpoints. First, we need to create a model with multiple containers as a parameter, followed by creating an endpoint configuration, and finally creating the endpoint.

Inference pipelines

We can also host real-time endpoints consisting of two to five containers as an inference pipeline behind a single endpoint. Each of these containers can be a pretrained SageMaker built-in algorithm, our custom algorithm, preprocessing code, predictions, or postprocessing code. All the containers in the inference pipeline function in a sequential manner. The first container processes the initial HTTP request. The response from the first container is sent as a request to the second container, and so on. The response from the final container is sent by SageMaker to the client. An inference pipeline can be considered as a single model that can be hosted behind a single endpoint or can also be used to run batch transform jobs. Since all the containers in an inference pipeline are running on the same EC2 instance, there is very low latency in communication between the containers.

Monitoring deployed models

After deploying a model into production, data scientists and machine learning engineers have to continuously check on the model's quality. This is because with time, the model's quality may drift and it may start to predict incorrectly. This may occur because of several reasons, such as a change in one or more variables' distribution in the dataset, the introduction of bias in the dataset with time, or some other unknown process or parameter being changed.

Traditionally, data scientists often run their analysis every few weeks to check if there has been any change in the data or model quality, and if there is, they go through the entire process of feature engineering, model training, and deployment again. With SageMaker Model Monitor, these steps can be automated. We can set up alarms to detect if there is any drift in data quality, model quality, bias drift in the model and feature distribution drift, and then take corrective actions such as fixing quality issues and retraining models.

Figure 7.15 shows an example of using SageMaker Model Monitor with an endpoint (real-time or asynchronous):

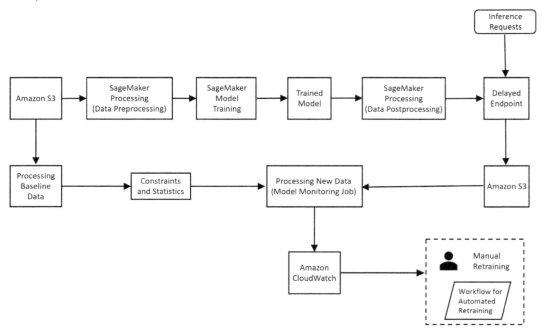

Figure 7.15 – SageMaker Model Monitor workflow for a deployed endpoint (real-time or asynchronous)

First, we need to enable Model Monitor on our endpoint, either at the time of creation or later. In addition, we also need to run a baseline processing job. This processing job analyzes the data and creates statistics and constraints for the baseline data (generally the same dataset with which the model has been trained or validated on). We also need to enable data capture on our SageMaker endpoint to be able to capture the new data along with inference results. Another processing job is then run on fixed intervals to create new statistics and constraints, compares them with the baseline statistics and constraints, and then configures alarms using Amazon CloudWatch metrics in case there is drift in any of the metrics we are analyzing.

In case of a violation in any of the metrics, alarms are generated and these can be sent to a data scientist or machine learning engineer for further analysis and model retraining if needed. Alternatively, we can also use these alarms to trigger a workflow (MLOps) to retrain our machine learning model. Model Monitor is a very valuable tool for data scientists. It can cut down on tedious manual processes for detecting bias and drift in data and model as time progresses after deploying a model.

Asynchronous inference

SageMaker real-time endpoints are suitable for machine learning use cases that have very low latency inference requirements (up to 60 seconds), along with the data size for inference not being large (maximum 6 MB). On the other hand, batch transforms are suitable for offline inference on very large datasets. Asynchronous inference is another relatively new inference option in SageMaker that can process data up to 1 GB and can take up to 15 minutes in processing inference requests. Hence, they are useful for use cases that do not have very low latency inference requirements.

Asynchronous endpoints have several similarities to real-time endpoints. To create asynchronous endpoints, like with real-time endpoints, we need to carry out the following steps:

1. Create a model.
2. Create an endpoint configuration for the asynchronous endpoint. There are some additional parameters for asynchronous endpoints.
3. Create the asynchronous endpoint.

Asynchronous endpoints also have differences when compared to real-time endpoints, as outlined here:

*   One main difference from real-time endpoints is that we can scale endpoint instances down to zero when there are no inference requests. This can cut down on the costs associated with having an endpoint for inference.
*   Another key difference compared to a real-time endpoint is that instead of passing the payload in line with the request for inference, we upload the data to an Amazon S3 location, and pass on the S3 URI along with the request. Internally, SageMaker keeps a queue of these requests and processes them in the order that they were received. Just like real-time endpoints, we can also do monitoring on the asynchronous endpoint in order to detect model and data drift, as well as new bias.

The code to show a SageMaker asynchronous endpoint is shown in *Figure 7.16*, using the same model that we created in the batch transform example.

```
[17]: #Defining configuration for asynchronous endpoint
      from sagemaker.async_inference import AsyncInferenceConfig
      async_config = AsyncInferenceConfig
```

```
[18]: async_config = AsyncInferenceConfig(
          output_path='s3://{}/{}/output'.format(bucket, prefix),
          max_concurrent_invocations_per_instance=10
      )
```

```
[19]: #Deploying the XGBoost model as an asynchronous endpoint
      async_predictor = xgb_model.deploy(async_inference_config=
                        async_config,initial_instance_count=1,
                        instance_type='ml.m4.xlarge')
```

Figure 7.16 – Creating an asynchronous endpoint configuration,
followed by the creation of the asynchronous endpoint

Figure 7.17 shows a sample of results for carrying out asynchronous inference on Abalone data used for the batch transform and real-time endpoints examples in previous sections.

```
[20]: #Carrying out inference using the asynchronous endpoint
      async_predictor.serializer = sagemaker.serializers.CSVSerializer()
      predictions = inference(test_data.drop(['Rings'], axis=1)
                              .to_numpy(), async_predictor)
      predictions = np.round(predictions).astype(int)
```

```
[29]: #Forming data frame with actual values and predictions and
      #displaying the head of the results
      df_output_realtime = pd.DataFrame({'Actual': test_data['Rings'],
                                         'Predictions': predictions})
      df_output_realtime.tail()
```

Figure 7.17 – Serializing the inference request and calling the
asynchronous endpoint to carry out inference on the data.

Figure 7.18 shows the actual and predicted results for a few data points.

Actual	Predictions
9	11
19	15
7	9
12	9
8	8

Figure 7.18 – Showing a few predicted results versus the actual
values (Rings) using the asynchronous endpoint

In the following section, we will look into the high availability and fault tolerance capabilities of SageMaker endpoints.

The high availability of model endpoints

Amazon SageMaker provides fault tolerance and high availability of the deployed endpoints. In this section, we will discuss various features and options of AWS cloud infrastructure and Amazon SageMaker, that we can use to ensure that our endpoints are fault-tolerant, resilient, and highly available.

Deployment on multiple instances

SageMaker gives us the option of deploying our endpoints on multiple instances. This protects from instance failures. If one instance goes down, then other instances can still serve the inference requests. In addition, if our endpoints are deployed on multiple instances and an availability zone outage occurs or an instance fails, SageMaker automatically tries to distribute our instances across different availability zones, thereby improving the resiliency of our endpoints. It is also a good practice to deploy our endpoints using small instance types spread across different availability zones.

Endpoints autoscaling

Oftentimes, we design our online applications for peak load and traffic. This is also true for machine learning-based applications, where we need hosted endpoints to carry out inference in real time or near real time. In such a scenario, we generally deploy models using the maximum number of instances in order to serve the peak workload and traffic.

If we don't do this, then our application may start timing out when there are more inference requests than the instances can handle in a combined fashion. This approach results in either the wastage of compute resources or interruptions in the service of our machine learning applications.

To avoid this kind of scenario, Amazon SageMaker lets us configure our endpoints with an autoscaling policy. Using the autoscaling policy, the number of instances on which our endpoint is deployed increases as traffic increases, and decreases as traffic decreases. This helps not only with the high availability of our inference endpoints but also helps in reducing the cost significantly. We can enable autoscaling for a model using either the SageMaker console, the AWS CLI, or the Application Auto Scaling API.

In order to apply autoscaling, we need an autoscaling policy. The autoscaling policy uses Amazon CloudWatch metrics and target values assigned by us to decide when to scale the instances up or down. We also need to define the minimum and maximum number of instances that the endpoint can be deployed on. Other components of the autoscaling policy include the required permissions to carry out autoscaling, a service-linked IAM role, and a cool-down period to wait after a scaling activity before starting the next scaling activity.

Figures 7.19 – 7.22 show autoscaling being configured on an asynchronous endpoint using the Amazon SageMaker console:

1. *Figure 7.19* shows that initially the endpoint is just deployed on a single instance:

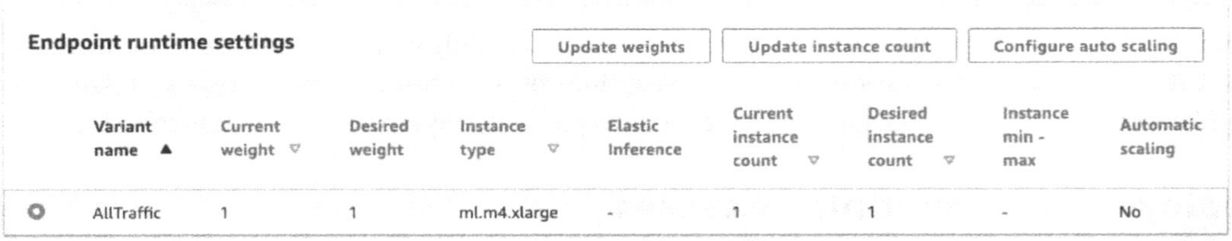

Endpoint runtime settings					Update weights	Update instance count	Configure auto scaling	
Variant name ▲	Current weight ▽	Desired weight	Instance type ▽	Elastic Inference	Current instance count ▽	Desired instance count ▽	Instance min - max	Automatic scaling
○ AllTraffic	1	1	ml.m4.xlarge	-	1	1	-	No

Figure 7.19 – The endpoint run time settings showing the endpoint being
deployed on an instance and autoscaling not being used

After clicking on the **Configure auto scaling** button, your screen should look like those shown in *Figure 7.20* and *Figure 7.21*.

2. As seen in *Figure 7.20*, we need to update the minimum and maximum instance counts to 2 and 10, respectively.

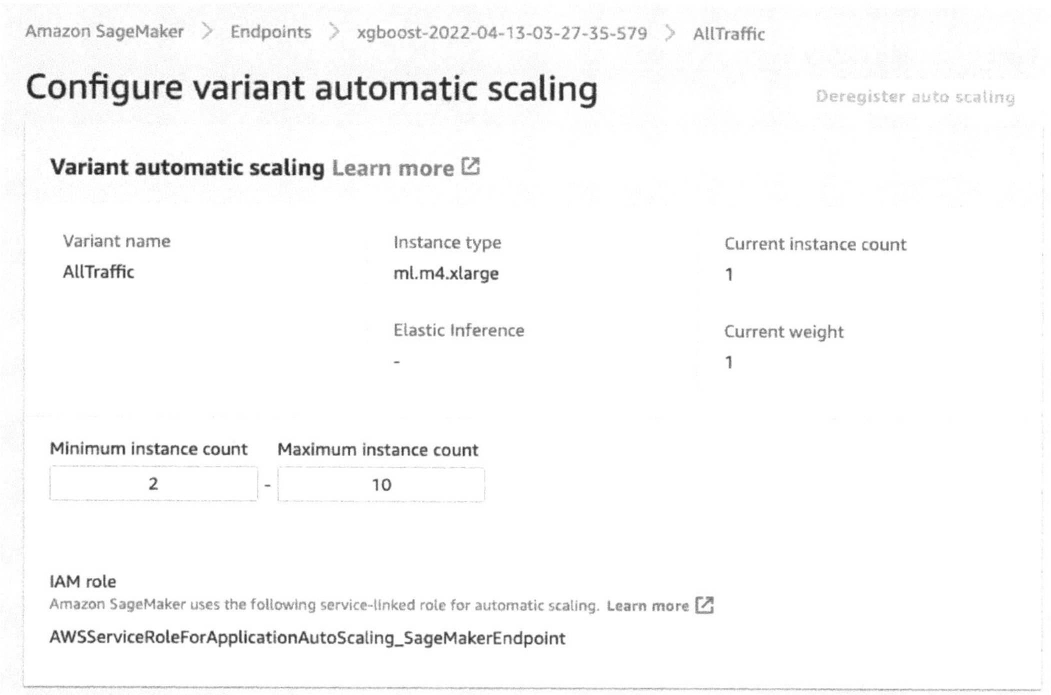

Amazon SageMaker > Endpoints > xgboost-2022-04-13-03-27-35-579 > AllTraffic

Configure variant automatic scaling

Deregister auto scaling

Variant automatic scaling Learn more ☑

Variant name	Instance type	Current instance count
AllTraffic	ml.m4.xlarge	1
	Elastic Inference	Current weight
	-	1

Minimum instance count   Maximum instance count

| 2 | - | 10 |

IAM role

Amazon SageMaker uses the following service-linked role for automatic scaling. Learn more ☑

AWSServiceRoleForApplicationAutoScaling_SageMakerEndpoint

Figure 7.20 – Configuring autoscaling on the endpoint to use 2 – 10
instances, depending on traffic (using SageMaker console).

3. As seen in *Figure 7.21*, we need to update the value for the **SageMakerVariantInvocationsPerInstance**
built-in metric to 200.

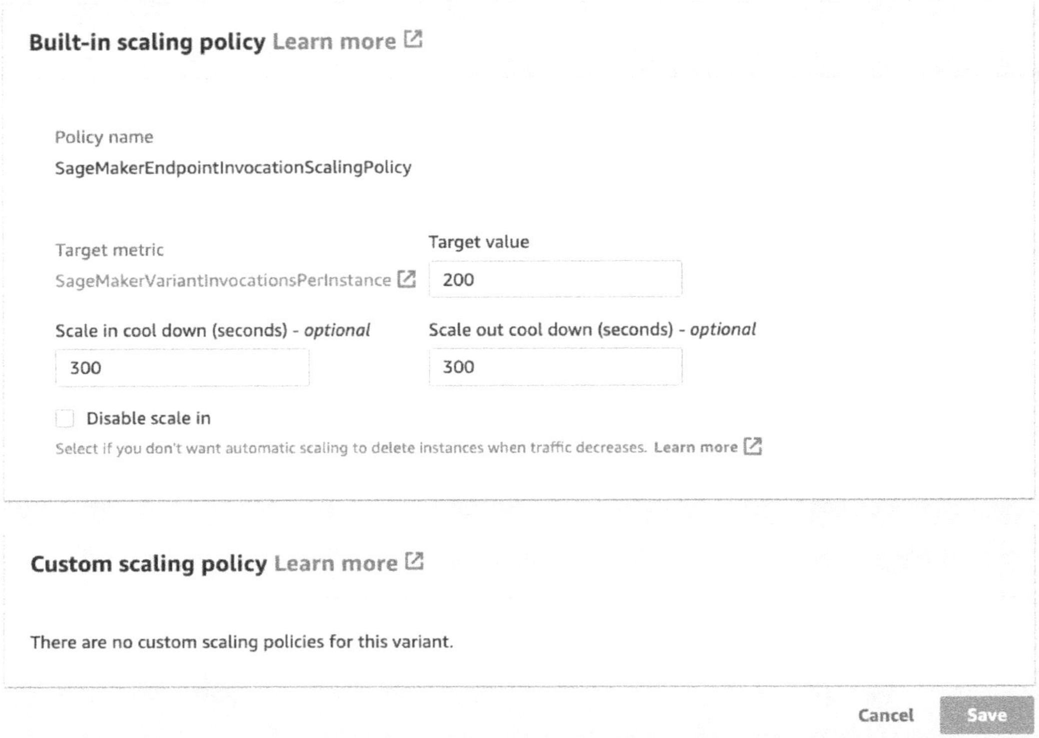

Built-in scaling policy Learn more ☒

Policy name

SageMakerEndpointInvocationScalingPolicy

Target metric

SageMakerVariantInvocationsPerInstance ☒

Target value

200

Scale in cool down (seconds) - *optional*

300

Scale out cool down (seconds) - *optional*

300

☐ Disable scale in

Select if you don't want automatic scaling to delete instances when traffic decreases. **Learn more** ☒

Custom scaling policy Learn more ☒

There are no custom scaling policies for this variant.

Cancel   Save

Figure 7.21 – Setting the scale in and out time along with the target metric threshold to trigger autoscaling

Once this target value is hit and we are out of the cool-down period, SageMaker will automatically start a new instance or shut down the instance regardless of whether we are above or below the target value.

4. *Figure 7.22* shows the results of the endpoint being updated with our autoscaling policy.

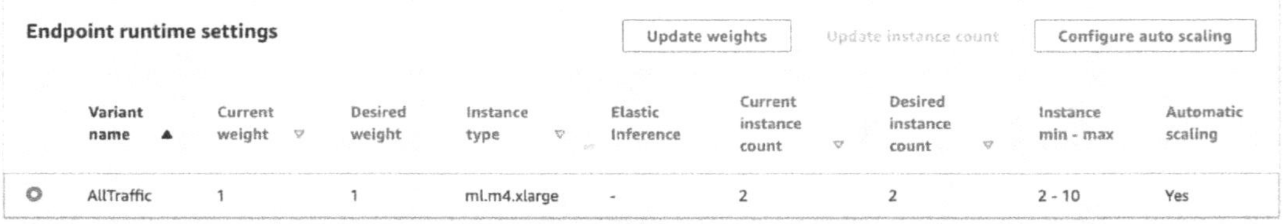

Endpoint runtime settings

Update weights   Update instance count   Configure auto scaling

Variant name ▲	Current weight ▽	Desired weight	Instance type ▽	Elastic Inference	Current instance count ▽	Desired instance count ▽	Instance min - max	Automatic scaling
⊙ AllTraffic	1	1	ml.m4.xlarge	-	2	2	2 - 10	Yes

Figure 7.22 – The endpoint running with the autoscaling policy applied

The autoscaling policy can also be updated or deleted after being applied.

Endpoint modification without disruption

We can also modify deployed endpoints without affecting the availability of the models deployed in production. In addition to applying an autoscaling policy as discussed in the previous section, we can also update compute instance configurations of the endpoints. We can also add new model variants and change the traffic patterns between different model variants. All these tasks can be achieved without negatively affecting the endpoints deployed in production.

Now that we've discussed how we can ensure that our endpoints are highly available, let's discuss Blue/green deployments next.

Blue/green deployments

In a production environment where our models are running to make inferences in real time or near real time, it is very important that when we need to update our endpoints that it can happen without any disruption or problems. Amazon SageMaker automatically uses blue/green deployment methodology whenever we update our endpoints. In this kind of scenario, a new fleet, called the green fleet, is provisioned with our updated endpoints. The workload is then shifted from the old fleet, called the blue fleet, to the green fleet. After an evaluation period to make sure that everything is running without any issues, the blue fleet is terminated. SageMaker also provides the following three different traffic-shifted modes for blue/green deployment, allowing us to have more control over the traffic-shifting patterns.

All at once

In this traffic-shifting mode, all of the traffic is shifted at once from the blue fleet to the green fleet. The blue (old) fleet is kept in service for a period of time (called the baking period) to make sure everything is working fine, and performance and functionality are as expected. After the baking period, the blue fleet is terminated all at once. This type of blue/green deployment minimizes the overall update duration, while also minimizing the cost. One disadvantage of the all-at-once method is that regressive updates affect all of the traffic since the entire traffic is shifted to the green fleet at once.

Canary

In canary blue/green deployment, a small portion of the traffic is first shifted from the blue fleet to the green fleet. This portion of the green fleet that starts serving a portion of the traffic is called the canary, and it should be less than 50% of the new fleet's capacity. If everything works fine during the baking period and no CloudWatch alarms are triggered, the rest of the traffic is also shifted to the green fleet, after which the blue fleet is terminated. If any alarms go off during the baking period, SageMaker rolls back all the traffic to the blue fleet.

An advantage of using canary blue/green deployment is that it confines the blast radius of regressive updates only to the canary fleet and not to the whole fleet, unlike all-at-once blue/green deployment. A disadvantage of using canary deployment is that both the blue and the green fleets are operational for the entire deployment period, thus adding to the cost.

Linear

In linear blue/green deployment, traffic is shifted from the blue fleet to the green fleet in a fixed number of pre-specified steps. In the beginning, the first portion of traffic is shifted to the green fleet, and the same portion of the blue fleet is deactivated. If no alarms go off during the baking period, SageMaker initiates the shifting of the second portion, and so on. If at any point an alarm goes off, SageMaker rolls back all the traffic to the blue fleet. Since traffic is shifted over to the green fleet in several steps, linear blue/green deployment reduces the risk of regressive updates significantly. The cost of the linear deployment method is proportional to the number of steps configured to shift the traffic from the blue fleet to the green fleet.

As discussed in this section, SageMaker has these blue/green deployment methods to make sure that there are safety guardrails when we are updating our endpoints. These blue/green deployment methods ensure that we are able to update our inference endpoints with no or minimal disruption to our deployed machine learning models in a production environment.

Let's now summarize what we've covered in this chapter.

Summary

In this chapter, we discussed the various managed deployment methods available when using Amazon SageMaker. We talked about the suitability of the different deployment/inference methods for different use case types. We showed examples of how we can do batch inference and deploy real-time and asynchronous endpoints. We also discussed how SageMaker can be configured to automatically scale both up and down, and how SageMaker ensures that in case of an outage, our endpoints are deployed to multiple availability zones. We also touched upon the various blue/green deployment methodologies available with Amazon SageMaker, in order to update our endpoints in production.

In a lot of real-world scenarios, we do not have high-performance clusters of instances available for carrying out inference on new and unseen data in real time. For such applications, we need to use edge computing devices. These devices often have limitations on compute power, memory, connectivity, and bandwidth, and need the models to be optimized to be able to use on these edge devices.

In the next chapter, we will extend this discussion to learn about using machine learning models on edge devices.

References

You can refer to the following resources for more information:

- Hosting multiple models on a single endpoint: `https://docs.aws.amazon.com/sagemaker/latest/dg/multi-model-endpoints.html`

- Amazon EC2 Inf1 instances: `https://aws.amazon.com/ec2/instance-types/inf1/`

- Using your own inference code with SageMaker Batch Transform: `https://docs.aws.amazon.com/sagemaker/latest/dg/your-algorithms-batch-code.html`

- Multiple models with different containers behind a single endpoint: `https://docs.aws.amazon.com/sagemaker/latest/dg/multi-container-endpoints.html`

- Serverless inference on Amazon SageMaker: `https://docs.aws.amazon.com/sagemaker/latest/dg/serverless-endpoints.html`

- Blue/green deployments using Amazon SageMaker: `https://docs.aws.amazon.com/sagemaker/latest/dg/deployment-guardrails-blue-green.html`

- Canary traffic shifting: `https://docs.aws.amazon.com/sagemaker/latest/dg/deployment-guardrails-blue-green-canary.html`

- SageMaker's Transformer class: `https://sagemaker.readthedocs.io/en/stable/api/inference/transformer.html?highlight=transformer`

- Amazon SageMaker real-time inference: `https://docs.aws.amazon.com/sagemaker/latest/dg/realtime-endpoints-deployment.html`

- Best practices for hosting models using Amazon SageMaker: `https://docs.aws.amazon.com/sagemaker/latest/dg/deployment-best-practices.html`

Optimizing and Managing Machine Learning Models for Edge Deployment

Every **Machine Learning** (**ML**) practitioner knows that the ML development life cycle is an extremely iterative process, from gathering, exploring, and engineering the right features for our algorithm, to training, tuning, and optimizing the ML model for deployment. As ML practitioners, we spend up to 80% of our time getting the right data for training the ML model, with the last 20% actually training and tuning the ML model. By the end of the process, we are all probably so relieved that we finally have an optimized ML model that we often don't pay enough attention to exactly how the resultant model is deployed. It is, therefore, important to realize that where and how the trained model gets deployed has a significant impact on the overall ML use case. For example, let's say that our ML use case was specific to **Autonomous Vehicles** (**AVs**), specifically a **Computer Vision** (**CV**) model that was trained to detect other vehicles. Once our CV model has been trained and optimized, we can deploy it.

But where do we deploy it? Do we deploy it on the vehicle itself, or do we deploy it on the same infrastructure used to train the model?

Well, if we deploy the model onto the same infrastructure we used to train the model, we will also need to ensure that the vehicle can connect to this infrastructure. We will also need to ensure that the connectivity from the vehicle to the model is sufficiently robust and performant to ensure that model inference results are timely. It would be disastrous if the vehicle was unable to detect an oncoming vehicle in time. So, in this use case, it might be better to execute model inferences on the vehicle itself; that way, we won't need to worry about network connectivity, resilience, bandwidth, and latency between the vehicle and the ML model. Essentially, deploying and executing ML model inferences on the edge devices.

However, deploying and managing ML models at the edge imposes additional complexities on the ML development life cycle. So, in this chapter, we will be reviewing some of these complexities, and by means of a practical example, we will see exactly how to optimize, manage, and deploy the ML model for the edge. Thus, we will be covering the following topics:

- Understanding edge computing
- Reviewing the key considerations for optimal edge deployments
- Designing an architecture for optimal edge deployments

Technical requirements

To work through the hands-on examples within this chapter, you should have the following prerequisites:

- A web browser (for the best experience, it is recommended that you use a Chrome or Firefox browser)
- Access to the AWS account that you've used in *Chapter 5, Data Analysis*
- Access to the Amazon SageMaker Studio development environment that we created in *Chapter 5, Data Analysis*
- Example code for this chapter is provided in the companion GitHub repository (`https://github.com/PacktPublishing/Applied-Machine-Learning-and-High-Performance-Computing-on-AWS/tree/main/Chapter08`)

Understanding edge computing

To understand how we can optimize, manage, and deploy ML models for the edge, we need to first understand what edge computing is. Edge computing is a pattern or type of architecture that brings data storage mechanisms, and computing resources closer to the actual source of the data. So, by bringing these resources closer to the data itself, we are fundamentally improving the responsiveness of the overall application and removing the requirement to provide optimal and resilient network bandwidth.

Therefore, if we refer to the AV example highlighted at the outset of this chapter, by moving the CV model closer to the source of the data, basically the live camera feed, we are able to detect other vehicles in real time. Consequently, instead of having our application make a connection to the infrastructure that hosts the trained model, we send the camera feed to the ML model, retrieve the inferences, and finally, have the application take some action based on the results.

Now, using an edge computing architecture, we can send the camera feed directly to the trained CV model, running on the compute resources inside the vehicle itself, and have the application take some action based on the retrieved inference results in real time. Hence, by using an edge computing architecture, we have alleviated any unnecessary application latency introduced by having to connect to the infrastructure hosting the CV model. Subsequently, we have allowed the vehicle to react to other detected vehicles in real time. Additionally, we have removed the dependency on a resilient and optical network connection.

However, while the edge computing architecture provides improved application response times, the architecture itself also introduces additional complexities, especially related to its design and implementation. So, in the next section, we will review some of the key considerations that need to be accounted for when optimally deploying an ML model to the edge.

Reviewing the key considerations for optimal edge deployments

As we saw in the previous two chapters, there are several key factors that need to be taken into account when designing an appropriate architecture for training as well as deploying ML models at scale. In both these chapters, we also saw how Amazon SageMaker can be used to implement an effective ephemeral infrastructure for executing these tasks. Hence, in a later part of this chapter, we will also review how SageMaker can be used to deploy ML models to the edge at scale. Nonetheless, before we can dive into edge deployments with SageMaker, it is important to review some of the key factors that influence the successful deployment of an ML model at the edge:

- Efficiency
- Performance
- Reliability
- Security

While not all the mentioned factors may influence how an edge architecture is designed and may not be vital to the ML use case, it is important to at least consider them. So, let's start by examining the significance of efficiency within the edge architecture design.

Efficiency

Efficiency, by definition, is the ratio, or percentage, of output correlated with the input. When the ML model is deployed at the edge, it makes the execution closer to the application, as well as to the input data being used to generate the inferences. Therefore, we can say that deploying an ML model to the edge makes it efficient by default. However, this assumption is based on the fact that the ML model only provides inference results based on the input data and doesn't need to perform any preprocessing of the input data beforehand.

For example, if we refer to the CV model example, if the image data provided to the ML model had to be preprocessed, for instance, the images needed to be resized, or the image tensors needed to be normalized, then this preprocessing step introduces more work for the ML model. Therefore, this ML model isn't as efficient as one that just provides the inference result without any preprocessing.

So, when designing an architecture for edge deployments, we need to reduce the amount of unnecessary work being performed on the input data, to further streamline that inference result and therefore make the inference as efficient as possible.

Performance

Measuring performance is similar to measuring efficiency, except that it's not a ratio but rather a measurement of quality. Thus, when it comes to measuring the quality of an ML model deployed to the edge, we measure how quickly the model provides an inference result and how true the inference result is. So, just as with efficiency, having the ML model inference results closer to the data source does improve the overall performance, but there are trade-offs that are specific to the ML model use case that also need to be considered.

To illustrate using the CV use case example, we may have to compromise on the quality of the model's inference results by compressing, or pruning the model architecture, essentially making it smaller to fit into the limited memory and run on the limited processing capacity of an edge computing device. Additionally, while most CV algorithms require GPU resources for training, as well as inference, we may not be able to provide GPU resources to edge devices.

So, when designing an architecture for edge deployments, we need to anticipate what computing resources are available at the edge and explore how to refactor the trained ML model to ensure that it will fit on the edge device and provide the best inference results in the shortest amount of time.

Reliability

Depending on the ML use case and how the model's inference results are used, reliability may not be a crucial factor influencing the design of the edge architecture. For instance, if we consider the AV use case, having the ability to detect other vehicles in proximity is a matter of life and death for the passenger. Alternatively, not being able to use an ML model to predict future temperature fluctuations on a smart thermostat may be an inconvenience but not necessarily a critical factor influencing the design of the edge architecture.

Being able to detect as well as alert when a deployed ML model fails are crucial aspects of the overall reliability of the edge architecture. Thus, the ability to manage ML models at the edge becomes a crucial element of the overall reliability of the edge architecture. Other factors that may influence the reliability and manageability of the architecture are the communication technologies in use. These technologies may provide different levels of reliability and may require multiple different types. For example, in the AV use case, the vehicle may use cellular connectivity as the primary communication technology, and should this fail, a satellite link may be used as a backup.

So, when designing an architecture for edge deployments, reliability might not be a critical factor, but having the ability to manage models deployed to the edge is also essential to the overall scalability of the architecture.

Security

As is the case with reliability, security may not be a crucial factor influencing the edge architecture's design, and more specific to the use case itself. For example, it may be necessary to encrypt all data stored on the edge architecture. With respect to the ML model deployed on the edge architecture, this means that all inference data (both requests and response data to and from the ML model) must be encrypted if persisted on the edge architecture. Additionally, any data transmitted between components internally and externally to the architecture must be encrypted as well.

It is important to bear in mind that there is a shift from a security management perspective from a centralized to a decentralized trust model and that compute resources within the edge architecture are constrained by size and performance capabilities. Consequently, the choice in the types of encryption is limited, as advanced encryption requires additional compute resources.

Now that we have reviewed some of the key factors that influence the design of an optimal edge architecture for ML model deployment, in the next section, we will dive into building an optimal edge architecture using Amazon SageMaker, as well as other AWS services that specialize in edge device management.

Designing an architecture for optimal edge deployments

While there are a number of key factors that influence the edge architecture design, as was highlighted in the previous section, there is also a critical capability necessary to enable these factors, namely the ability to build, deploy, and manage the device software at the edge. Additionally, we also need the ability to manage the application, in essence, the ML model deployed to run on the edge devices. Consequently, AWS provides both of these management capabilities using a dedicated device management service called **AWS IoT Greengrass** (`https://aws.amazon.com/greengrass/`), as well as the ML model management capability built into Amazon SageMaker called **Amazon SageMaker Edge** (`https://aws.amazon.com/sagemaker/edge`). AWS IoT Greengrass is a service provided by AWS to deploy software to remote devices at scale without firmware updates. *Figure 8.1* shows an example of an architecture that leverages both Greengrass and SageMaker to support edge deployments.

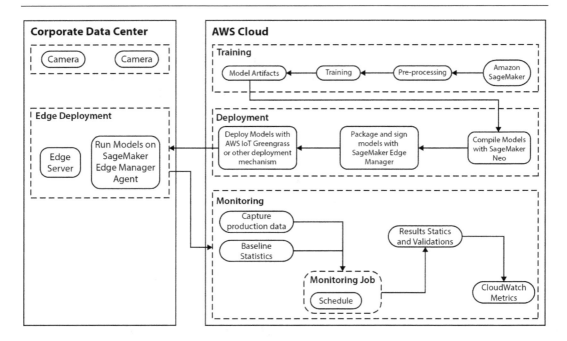

Figure 8.1 – Architecture for edge deployments using Amazon SageMaker and IoT Greengrass

As you can see from *Figure 8.1*, a typical architecture is divided into two separate and individual architectures. One for the cloud-based components and one for the edge components. At a high level, the cloud environment is used to build, deploy, and manage the use case application being deployed to the edge. The corresponding edge environment, or the corporate data center, in this case, is where the edge devices reside, and in turn, execute upon the ML use case by running the supported ML models. From an ML use case perspective, *Figure 8.1* also shows the cameras attached to the edge server device, allowing any video captured to be streamed to the ML model in order for the model to classify the objects detected in the video frames. *Figure 8.1* shows a simplistic flow for the CV use case, from training an ML model on SageMaker to deploying it to the edge server using Greengrass, and then managing and monitoring the solution. But building out the actual solution is very complicated.

So, to illustrate this complexity, in the next section, we are going to build out this architecture by breaking out each component, starting with the corporate data center or edge architecture.

Building the edge components

As highlighted in the previous section, the corporate data center serves as the edge location for our CV use case. Inside this edge location, we have a number of cameras, connected to a compute device, or edge server, that runs the CV model. Since we don't have access to a corporate data center, within the context of the book, we will simulate building out the edge environment using an **Elastic Compute Cloud (EC2)** instance.

> **Note**
>
> If you are not accustomed to working with EC2 instances, you can familiarize yourself with them by referencing the following documentation: `https://aws.amazon.com/ec2/getting-started/`.

The following steps will demonstrate how to set up the edge server using an EC2 instance and configure the necessary Greengrass software as well as the required security permissions. Let's get started with setting up the appropriate **Identity and Access Management** (**IAM**) roles and permissions:

1. Log into your AWS account and open the IAM console (`https://console.aws.amazon.com/iam/home`).

2. Once the IAM console is open, use the left-hand navigation panel and click on **Roles** to open the **Roles** dashboard. Then click on the **Create role** button in the top-right corner.

3. Once the **Create role** wizard starts, select **AWS Service** as the **Trusted entity** type.

4. For the **Use case**, select **EC2** and click on the **Next** button.

5. On the **Add permissions** page, click on the **Create policy** button in the top right to open the **Create policy** page.

6. On the **Create policy** page, select the **JSON** tab, paste the policy from the `SageMakerGreenGrassV2MinimalResourcePolicy.json` file in GitHub (`https://github.com/PacktPublishing/Applied-Machine-Learning-and-High-Performance-Computing-on-AWS/blob/main/Chapter08/SageMakerGreenGrassV2MinimalResourcePolicy.json`), and make sure to update the `<account_id>` tag in the policy with your AWS account ID.

7. Click on the **Next: Tags** button.

8. Click on the **Next: Review** button.

9. On the **Review policy** page, name the policy `SageMakerGreenGrassV2MinimalResourcePolicy`, and click the **Create policy** button.

10. Go back to the **Add permission** page from *step 5*, and refresh the page to capture the newly created IAM policy.

11. Search for the `SageMakerGreenGrassV2MinimalResourcePolicy` policy in the search bar, and once found, select the checkbox for the policy to add the permission, then click the **Next** button.

12. On the **Role details** page, enter `SageMakerGreenGrassV2MinimalResourceRole`, as the **Role name**.

13. Once again, using the **Role** dashboard, click the newly created `SageMakerGreenGrassV2MinimalResourceRole` role to open the **Role summary** page.

14. Now, click the **Add permissions** dropdown and select the **Attach policies** option.

15. Search for the `AmazonSageMakerEdgeDeviceFleetPolicy` policy, and click on the checkbox to select this policy.

16. Repeat the process shown in *step 15*, except this time, select the checkbox for the **AmazonSageMakerFullAccess**, **AWSIoTLogging**, **AmazonS3FullAccess**, **AWSIoTRuleActions**, **AWSIoTThingsRegistration**, and **AmazonSSMManagedInstanceCore** policies.

17. With these policies selected, click the **Attach policies** button.

18. In the **Role summary** page from *step 13*, click the **Add permissions** dropdown and select the **Create inline policy** option.

19. On the **Create policy** page, click the **JSON** tab and add the following policy statement:

```
{
    "Version":"2012-10-17",
    "Statement":[
        {
            "Sid":"GreengrassComponentAccess",
            "Effect":"Allow",
            "Action":[
                "greengrass:CreateComponentVersion",
                "greengrass:DescribeComponent"
            ],
            "Resource":"*"
        }
    ]
}
```

20. Click the **Review policy** button.

21. Name the policy `GreengrassComponentAccessPolicy` and click the **Create policy** button.

22. In the **Role summary** page, click the **Trust relationships** tab and click the **Edit trust policy** button.

23. Replace the existing policy with the following trust policy statement:

```
{
    "Version": "2012-10-17",
    "Statement": [
        {
            "Effect": "Allow",
            "Principal": {
                "Service": "ec2.amazonaws.com"
```

```
            },
            "Action": "sts:AssumeRole"
        },
        {
            "Effect": "Allow",
            "Principal": {
                "Service": "credentials.iot.amazonaws.
    com"
            },
            "Action": "sts:AssumeRole"
        },
        {
            "Effect": "Allow",
            "Principal": {
                "Service": "sagemaker.amazonaws.com"
            },
            "Action": "sts:AssumeRole"
        }
    ]
}
```

Now that we have set up the necessary permissions, next we can configure the edge server with the following steps showing us how to configure the EC2 instance:

1. Before creating the EC2 instance, we need to configure the necessary scripts that customize the EC2 instance as an edge server. To provide easy command line access to AWS resources, open the **AWS CloudShell** console (https://console.aws.amazon.com/cloudshell/home).

2. Once the CloudShell console has been initialized in the browser, clone the companion GitHub repository by running the following command:

   ```
   $ git clone https://github.com/PacktPublishing/Applied-
   Machine-Learning-and-High-Performance-Computing-on-AWS
   src && cd src/Chapter08
   ```

3. Next, we create an S3 bucket and store the configuration scripts for the EC2 instance by running the following commands:

   ```
   $ export AWS_ACCOUNT=$(aws sts get-caller-identity
   --query "Account" --output text)
   ```

```
$ aws s3 mb s3://ec2-scripts-$AWS_REGION-$AWS_ACCOUNT
$ aws s3 sync scripts s3://ec2-scripts-$AWS_REGION-$AWS_
ACCOUNT/scripts
```

4. Run the following command to capture the name of the S3 bucket containing the EC2 configuration scripts:

```
$ echo ec2-scripts-$AWS_REGION-$AWS_ACCOUNT
```

> **Note**
>
> Make sure to remember the name of the S3 bucket containing the EC2 instance configuration scripts, as it will be used in a later step.

5. Open the EC2 management console (https://console.aws.amazon.com/ec2/v2/home) in a browser tab.

6. Once the EC2 console is open, click the **Launch instance** button.

7. In the **Launch an instance** wizard, in the **Name and tags** section, enter edge-server as the name for this instance.

8. Select AMI as **amzn2-ami-kernel-5.10-hvm-2.0.20220426.0-x86_64-gp2** and instance type as **c5.large**.

9. Scroll down to the **Configure storage** section and specify 20 GiB for **Root volume**.

10. Using the **IAM instance profile** drop-down box in the **Advance details** section, select the **SageMakerGreenGrassV2MinimalResourceRole** role.

11. In the **User data** text box of the **Advance details** section, paste the following Bash code:

```
#!/bin/bash
aws s3 cp s3://<REPLACE WITH THE NAME OF YOUR S3 BUCKET>/
scripts/ /home/ec2-user --recursive
sleep 30
process_id=$!
wait $process_id
sudo yum update -y
sudo yum install docker -y
sudo yum install python-pip -y
sudo pip3 install boto3
sudo pip3 install requests
cd /home/ec2-user
python3 getResourceTempCredentials.py
```

```
sudo service docker start
sudo usermod -a -G docker ec2-user
docker build -t "aws-iot-greensgrass:2.5" ./
chmod +x dockerRun.sh
```

> **Note**
>
> Make sure to replace the S3 bucket name with the name of your S3 bucket from the output of *step 4.*

12. Click the **Launch instance** button to create the EC2 instance.

13. Wait 10 minutes after the EC2 instance is in the active state before logging in to the EC2 instance, as the Bash script in the user data will need some time to install the necessary packages and build the Docker image for AWS IoT Greengrass software.

14. Log into the EC2 instance, and ensure you are in the /home/ec2-user directory. Using **vi**, open the env file, and ensure the AWS_REGION variable is set to the current AWS Region being used. For example, the following output shows the env file configured for the us-east-1 Region:

```
GGC_ROOT_PATH=/greengrass/v2
AWS_REGION=us-east-1
PROVISION=true
THING_NAME=mything
THING_GROUP_NAME=mythinggroupname
TES_ROLE_NAME=GreengrassV2TokenExchangeRole
TES_ROLE_ALIAS_NAME=GreengrassCoreTokenExchangeRoleAlias
COMPONENT_DEFAULT_USER=ggc_user:ggc_group
DEPLOY_DEV_TOOLS=true
```

> **Note**
>
> You can also customize the THING_NAME and THING_GROUP_NAME parameters. However, make sure these variables are in lowercase.

15. Save and exit the env file.

16. Run the Docker container by executing the following command:

```
$ ./dockerRun.sh
```

> **Note**
>
> In case you need to restart the Greengrass V2 Docker container, make sure to get the new credentials by running the following command:
>
> ```
> $ rm credentials #remove the old credentials file
> $ sudo python3 getResourceTempCredentials.py # create credentials
> file with new credentials
> $ ./dockerRun.sh
> ```

17. Open another terminal window, log in to the EC2 instance, and run the following command to verify that Greengrass V2 is running and retrieve the container ID:

    ```
    $ docker ps
    ```

18. You can then run the following command to access the container and explore the AWS IoT Greengrass Core software running inside the container:

    ```
    $ docker exec -it container-id /bin/bash
    ```

> **Note**
>
> When you use docker exec to run commands inside the Docker container, these are not captured in the Docker logs. As a best practice, we recommend you log your commands in the Docker logs so that you can look into the state of the Greengrass Docker container in case you need to troubleshoot any issues.

19. Run the following command in a different terminal. It will attach your terminal's input, output, and error to the container running currently. This will help you to view and control the container from your terminal.

    ```
    $ docker attach container-id
    ```

By executing these steps, we have effectively configured an EC2 instance to run our ML model as an edge server. So, with the edge components of our architecture successfully built, we can move on to building the ML model.

Building the ML model

Building the ML model involves training an optimal ML model to suit our business use case. As we've seen in previous chapters, Amazon SageMaker provides us with distinct capabilities that allow us to ingest and process the necessary training data, as well as train and optimize the best ML model. Additionally, we saw that SageMaker also allows us to deploy and host these models in the cloud, and as we will see, SageMaker also allows us to deploy and manage ML models at the edge.

The following steps will walk you through how to build an ML model that suits our use case, compile the model for an edge environment, and then deploy the model to the edge, all using SageMaker:

1. Within your AWS account, open the Amazon SageMaker management console (`https://console.aws.amazon.com/sagemaker/home`).

2. Launch the SageMaker Studio IDE.

> **Note**
>
> For more information on how to create and launch the SageMaker Studio IDE, please refer to the *Setting up EMR and SageMaker Studio* section of *Chapter 5, Data Analysis*.

3. Once the SageMaker Studio IDE has launched, use the **File Browser** navigation panel and double-click on the cloned `Applied-Machine-Learning-and-High-Performance-Computing-on-AWS` folder to expand it.

4. Then, double-click on the `Chapter_8` folder to open it for browsing.

5. Double-click on the `sagemaker_notebook` folder, and then launch the `1_compile_resnet_model_egde_manager.ipynb` notebook.

6. Once the notebook has started, use the menu bar to select the **Kernel** menu and then the **Restart Kernel and Run All Cells…** option.

Once the notebook has run, we will have our image classification model running on the edge server and managed as part of a fleet. Nonetheless, let's verify this by reviewing some of the important code cells in the notebook. The first part of the notebook downloads an already optimized or pre-trained **ResNet-18** (`https://arxiv.org/pdf/1512.03385.pdf`) model. This is essentially our optimized ML for the image classification use case that we have deployed to the edge. After setting up the SageMaker environment variables and establishing the necessary permissions, we upload the pre-trained model to S3. Once the model has been stored in the cloud, and as you can see from the following code snippet, we instantiate the pre-trained ResNet-18 model as a PyTorch-based SageMaker model object called `sagemaker_model`:

```
from sagemaker.pytorch.model import PyTorchModel
from sagemaker.predictor import Predictor

sagemaker_model = PyTorchModel(
    model_data=model_uri,
    predictor_cls=Predictor,
    framework_version=framework_version,
    role=role,
    sagemaker_session=sagemaker_session,
```

```
        entry_point="inference.py",
        source_dir="code",
        py_version="py3",
        env={"MMS_DEFAULT_RESPONSE_TIMEOUT": "500"},
)
```

With the model object defined, we then use SageMaker Neo (https://docs.aws.amazon.
com/sagemaker/latest/dg/neo.html) to compile a model that suites the specific compute
architecture of the edge device, in this case, our X86_64 Linux edge server:

```
sagemaker_client = boto3.client("sagemaker", region_
name=region)
target_arch = "X86_64"
target_os = 'LINUX'
response = sagemaker_client.create_compilation_job(
    CompilationJobName=compilation_job_name,
    RoleArn=role,
    InputConfig={
        "S3Uri": sagemaker_model.model_data,
        "DataInputConfig": data_shape,
        "Framework": framework,
    },
    OutputConfig={
        "S3OutputLocation": compiled_model_path,
#        "TargetDevice": 'jetson_nano',
        "TargetPlatform": {
            "Arch": target_arch,
            "Os": target_os
        },
    },
    StoppingCondition={"MaxRuntimeInSeconds": 900},
)
```

> **Note**
> Make sure to take note of the S3 path for the compiled model, as we will be using this path to
> deploy the model package.

After compiling the model for the edge server, we verify the model's functionality by deploying it as a SageMaker-hosted endpoint, and then using the following code to generate a sample inference for a test image:

```python
import numpy as np
import json

with open("horse_cart.jpg", "rb") as f:
    payload = f.read()
    payload = bytearray(payload)

response = runtime.invoke_endpoint(
    EndpointName=ENDPOINT_NAME,
    ContentType='application/octet-stream',
    Body=payload,
    Accept = 'application/json')
result = response['Body'].read()
result = json.loads(result)
print(result)
```

Once we've verified that the model functions correctly, essentially being able to classify the image correctly, we can then package the model as shown in the following code snippet:

```python
packaging_job_name = compilation_job_name + "-packaging-ggv2"
component_name = "SagemakerEdgeManager" + packaging_job_name

response = sagemaker_client.create_edge_packaging_job(
    RoleArn=role,
    OutputConfig={
        "S3OutputLocation": s3_edge_output_location,
    },
    ModelName=packaged_model_name,
    ModelVersion=model_version,
    EdgePackagingJobName=packaging_job_name,
    CompilationJobName=compilation_job_name,
)
```

Finally, once the model has been packaged, we can use the `create_device_fleet()` method to create a manageable fleet of edge devices to host the newly compiled ML model and then use the `register_device()` method to initialize our EC2 edge server as a registered or managed edge device that runs our ML model:

```
s3_device_fleet_output = os.path.join(s3_edge_output_location,
'fleet')
iot_role_arn = f'arn:aws:iam::{account_id}:role/
SageMakerGreenGrassV2MinimalResourceRole'
device_fleet_name = "mydevicefleet"
device_name = 'mything'

sagemaker_client.create_device_fleet(
    DeviceFleetName=device_fleet_name,
    RoleArn=iot_role_arn, # IoT Role ARN specified in previous
step
    OutputConfig={
        'S3OutputLocation': s3_device_fleet_output
    }
)
sagemaker_client.register_devices(
    DeviceFleetName=device_fleet_name,
    Devices=[
        {
            "DeviceName": device_name,
            "IotThingName": device_name
        }
    ]
)
```

Once the model has been trained and compiled and the edge server registered as a SageMaker-managed edge device, we can go ahead and deploy the model package to the edge server.

Deploying the model package

To deploy the model package to the edge server, we will register it as a Greengrass component and then deploy the component to the edge server using the Greengrass console. The following steps will walk us through how to do this:

1. Using a web browser, open the AWS IoT management console (https://console.aws. amazon.com/iot/home).

2. In the **Manage** section of the left-hand navigation panel, expand the **Greengrass devices** option and click on **Components**.

3. Click on the **Create component** button.

4. After the **Create component** page loads, ensure that the **Enter recipe as JSON** option is selected, and paste the `com.greengrass.SageMakerEdgeManager.ImageClassification.Model.json` file contents into the **Recipe** box.

> **Note**
>
> The `com.greengrass.SageMakerEdgeManager.ImageClassification.Model.json` file can be found in the companion GitHub repository (`https://github.com/PacktPublishing/Applied-Machine-Learning-and-High-Performance-Computing-on-AWS/blob/main/Chapter08/custom_component_recipes/com.greengrass.SageMakerEdgeManager.ImageClassification.Model.json`).

5. Update the S3 location of the packaged model under the `Artifacts` tag of the JSON file to match the S3 path for the compiled model.

> **Note**
>
> The S3 location of the packaged model is the output from the `create_compilation_job()` method used in the previous section.

6. Click on the **Create component** button.

7. Go to your S3 bucket and create the `artifacts` folder, and inside it, create another folder with the name `com.greengrass.SageMakerEdgeManager.ImageClassification`. Your S3 path should look like this: `s3://<bucket_name>/artifacts/com.greengrass.SageMakerEdgeManager.ImageClassification/`.

8. Upload the `image_classification.zip` and `installer.sh` files from the GitHub repo (`https://github.com/PacktPublishing/Applied-Machine-Learning-and-High-Performance-Computing-on-AWS/tree/main/Chapter08`) to the S3 location defined in *step 7*.

9. Update the S3 location of the `image_classification.zip` and `installer.sh` files under the `Artifacts` tag of the JSON file to match the S3 path defined in *step 8*.

10. Repeat *step 3* and *step 4* for the **com.greengrass.SageMakerEdgeManager.ImageClassification** component.

> **Note**
>
> The com.greengrass.SageMakerEdgeManager.ImageClassification. json file can be found in the companion GitHub repository (https://github.com/ PacktPublishing/Applied-Machine-Learning-and-High-Performance- Computing-on-AWS/blob/main/Chapter08/custom_component_recipes/ com.greengrass.SageMakerEdgeManager.ImageClassification.json).

With the Greengrass components registered, we can now deploy them to the edge server and run image classification inference using the components that we just created. Deploying these components downloads a SageMaker Neo-compiled pre-trained ML model and installs the SageMaker Edge Manager agent on the edge server. However, before we can actually perform the deployment of these components, we have to subscribe to the notification topic in order to view and manage any inference tasks on the edge server. The following steps will walk us through how to subscribe to a topic:

1. Go to the AWS IoT console (https://console.aws.amazon.com/iot/) and click on **MQTT test client** from the **Test** option of the left-hand navigation panel.

2. In the **Topic name** box of the **Subscribe to a topic** tab, enter gg/sageMakerEdgeManager/ image-classification.

3. Click the **Subscribe** button.

With the ability to monitor and manage any inference requests to the ML model running on the edge server in place, we can deploy the Greengrass components. The following steps will show us how to do this (they are also highlighted in the AWS Greengrass Developer guide at https://docs.aws. amazon.com/greengrass/v2/developerguide/greengrass-v2-developer- guide.pdf):

1. In the AWS IoT Greengrass console (https://console.aws.amazon.com/greengrass) navigation menu, choose **Deployments**, and then choose the deployment for the target device.

2. On the deployment page, choose **Revise** and then choose **Revise deployment**.

3. On the **Specify target** page, click **Next**.

4. Under the **My components** option of the **Select components** page, select both the **com.greengrass.SageMakerEdgeManager.ImageClassification**, and **com.greengrass.SageMakerEdgeManager.ImageClassification.Model** components.

5. Under **Public components**, turn off the **Show only selected components** toggle, and then select the **aws.greengrass.SageMakerEdgeManager** component.

6. Click **Next**.

7. On the **Configure components** page, select the **aws.greengrass.SageMakerEdgeManager** component, and choose **Configure component**.

8. Under **Configuration update**, in **Configuration to merge**, enter the following configuration:

```
{
    "DeviceFleetName": "device-fleet-name",
    "BucketName": "S3-BUCKET"
}
```

> **Note**
>
> Replace device-fleet-name and the S3 bucket name with the corresponding values you created when creating the device fleet in the notebook.

9. Choose **Confirm**, and then choose **Next**.

10. On the **Configure advanced settings** page, keep the default configuration settings and choose **Next**.

11. On the **Review** page, choose **Deploy**.

The deployment can take several minutes to complete. After the components have been deployed, we can view, manage, and monitor the ML model inference results in the component log of the Greengrass Core device, as well as in the AWS IoT MQTT client of the AWS IoT console. To view the inference results in the component log of the Greengrass Core device, log into the edge server EC2 instance and run the following command:

```
$ sudo tail -f /greengrass/v2/logs/com.greengrass.
SageMakerEdgeManager.ImageClassification.log
```

You should see Top 5 predictions with score 0.3 or above in the logs similar to the one shown here:

```
2022-08-15T03:38:45.437Z [INFO] (Copier) com.greengrass.
SageMakerEdgeManager.ImageClassification: stdout. {"timestamp":
"2022-08-15 03:38:45.282879+00:00", "inference-type": "image-
classification", "inference-description": "Top 5 predictions
with score 0.3 or above ", "inference-results": [{"Label":
"slot, one-armed bandit", "Score": "83.41295623779297"},
{"Label": "mousetrap", "Score": "75.826416015625"},
{"Label": "comic book", "Score": "73.64051055908203"},
{"Label": "microphone, mike", "Score": "71.14073181152344"},
{"Label": "honeycomb", "Score": "68.3149185180664"}]}.
{scriptName=services.com.greengrass.SageMakerEdgeManager.
ImageClassification.lifecycle.run.script, serviceName=com.
greengrass.SageMakerEdgeManager.ImageClassification,
currentState=RUNNING}
```

Alternatively, you can also view the results in **MQTT test client** on the AWS IoT console (`https://console.aws.amazon.com/iot/`) by clicking **MQTT test client** from the **Test** option on the left-hand navigation panel. In the **Subscriptions** section, you will see the prediction results, as shown in the following screenshot:

Figure 8.2 – Inference results on MQTT test client on AWS IoT console

If you can't see inference results in the MQTT client, the deployment might have failed or it did not reach the core device. This can occur primarily due to two reasons: your core device is not connected to the network, or it doesn't have the right permissions to execute the component. To troubleshoot it, you can run the following command on your core device. This command will open the AWS IoT Greengrass Core software log file, which includes logs from the Greengrass Core device's deployment service.

```
$ sudo tail -f /greengrass/v2/logs/greengrass.log
```

> **Note**
>
> For more information, see the troubleshooting documentation: `https://docs.aws.amazon.com/greengrass/v2/developerguide/ml-troubleshooting.html`.

With the Greengrass components now deployed to the edge server, we have successfully deployed our ML model to the edge. Furthermore, by leveraging the capabilities of AWS IoT Greengrass, as well as Amazon SageMaker, we have not only compiled the ML to function *efficiently* and ensure *performance* on the edge device but also established a mechanism to *manage* and *monitor* the environment. As you will recall from the *Reviewing the key considerations for optimal edge deployments* section, these are the key factors that make up an optimal edge architecture.

Summary

In this chapter, we introduced you to the concept of deploying ML models outside of the cloud, primarily on an edge architecture. To lay the foundation for how to accomplish an edge deployment, we also examined what an edge architecture is, as well as the most important factors that need to be considered when designing an edge architecture, namely efficiency, performance, and reliability.

With these factors in mind, we explored how the AWS IoT Greengrass, as well as Amazon SageMaker services, can be used to build an optimal ML model package in the cloud, compiled to run efficiently on an edge device, and then deployed to the edge environment, in a reliable manner. In doing so, we also highlighted just how crucial the ability to manage and monitor both the edge devices, as well as the deployed ML models is to create an optimal edge architecture.

In the next chapter, we will continue along the lines of performance monitoring and optimization of deployed ML models.

9
Performance Optimization for Real-Time Inference

Machine Learning (ML) and **Deep Learning** (DL) models are used in almost every industry, such as e-commerce, manufacturing, life sciences, and finance. Due to this, there have been meaningful innovations to improve the performance of these models. Since the introduction of transformer-based models in 2018, which were initially developed for **Natural Language Processing** (NLP) applications, the size of the models and the datasets required to train the models has grown exponentially. **Transformer-based** models are now used for forecasting as well as **computer vision** applications, in addition to NLP.

Let's travel back in time a little to understand the growth in size of these models. **Embeddings from Language Models (ELMo)**, which was introduced in 2018, had *93.6 million parameters*, while the **Generative Pretrained Transformer** model (also known as **GPT-3**), in 2020, had *175 billion parameters*. Today, we have DL models such as **Switch Transformers** (`https://arxiv.org/pdf/2101.03961.pdf`) with more than *1 trillion parameters*. However, the speed of innovation of hardware to train and deploy such models is not catching up with the speed of innovation of large models. Therefore, we need sophisticated techniques to train and deploy these models in a cost-effective yet performant way.

One way to address this is to think in terms of reducing the memory footprint of the model. Moreover, many inference workloads must provide flexibility, high availability, and the ability to scale as enterprises serve millions or billions of users, especially for real-time or near real-time use cases. We need to understand which instance type and how many instances to use for deployment. We should also understand the key metrics based on which we will optimize the models.

Therefore, to collectively address the preceding scenarios, in this chapter, we will dive deep into the following topics:

- Reducing the memory footprint of DL models
- Key metrics for optimizing models
- Choosing instance type, load testing, and performance tuning for models
- Observing results

> **Important note**
>
> For details on training large DL models, please refer to *Chapter 6, Distributed Training of Machine Learning Models*, where we cover the topic in detail along with an example.

Technical requirements

You should have the following prerequisites before getting started with this chapter:

- A web browser (for the best experience, it is recommended that you use the Chrome or Firefox browser)

- Access to the AWS account that you used in *Chapter 5, Data Analysis*

- Access to the SageMaker Studio development environment that we created in *Chapter 5, Data Analysis*

- Example Jupyter notebooks for this chapter are provided in the companion GitHub repository (https://github.com/PacktPublishing/Applied-Machine-Learning-and-High-Performance-Computing-on-AWS/tree/main/Chapter09)

Reducing the memory footprint of DL models

Once we have trained the model, we need to deploy the model to get predictions, which are then used to provide business insights. Sometimes, our model can be bigger than the size of the single GPU memory available on the market today. In that case, you have two options – either to reduce the memory footprint of the model or use distributed deployment techniques. Therefore, in this section, we will discuss the following techniques to reduce the memory footprint of the model:

- Pruning
- Quantization
- Model compilation

Let's dive deeper into each of these techniques, starting with pruning.

Pruning

Pruning is the technique of eliminating weights and parameters within a DL model that have little or no impact on the performance of the model but a significant impact on the inference speed and size of the model. The idea behind pruning methods is to make the model's memory and power efficient, reducing the storage requirement and latency of the model. A DL model is basically a neural network with many hidden layers connected to each other. As the size of the model increases, the number of hidden layers, parameters, and weight connections between the layers also increases. Therefore, pruning methods tend to remove unused parameters and weight connections without too much bearing on the accuracy of the model, as shown in *Figure 9.1* and *Figure 9.2*. *Figure 9.1* shows a neural network before pruning:

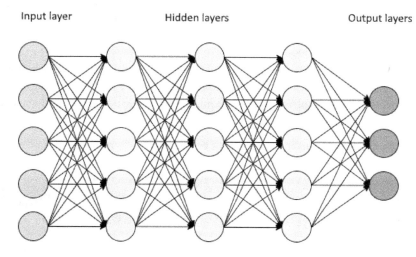

Figure 9.1 – Simple neural network before pruning

Figure 9.2 shows the same neural network after pruning:

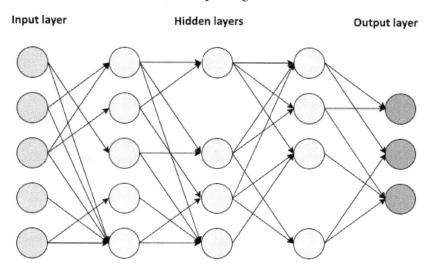

Figure 9.2 – Simple neural network after pruning

Now that we've covered pruning, let's take a look at quantization next.

Quantization

To train a neural network, data is first passed through the network in a forward pass, which calculates the activations, and then a backward pass, which uses the activations to calculate the gradients. The activations and gradients are usually stored in floating point 32, which takes 4 bytes of memory. When you have models with billions or trillions of parameters, this number is pretty significant. Therefore, **quantization** is a technique that reduces the model size by decreasing the precision of weights, biases, and activations of the model, such as floating point 16 or 8, or even to integer 8, which takes significantly less memory. For example, the GPT-J-6B model, which has 6 billion trainable parameters, takes about 23 GB of memory, as shown in *Figure 9.3*.

Figure 9.3 shows the number of parameters and size of the model when loaded from the Hugging Face library:

```python
from transformers import AutoModelForCausalLM

model = AutoModelForCausalLM.from_pretrained("EleutherAI/gpt-j-6B")

model.num_parameters()

6050882784

param_size = 0
buffer_size = 0
for param in model.parameters():
    param_size += param.nelement() * param.element_size()

for buffer in model.buffers():
    buffer_size += buffer.nelement() * buffer.element_size()

size_all_mb = (param_size + buffer_size) / 1024**2
print('Size: {:.3f} GB'.format(size_all_mb/1000))

Size: 23.194 GB
```

Figure 9.3 – The model size and number of parameters for the GPT-J-6B model

The same model when loaded with 16-bit floating point precision takes about 11 GB of memory, which is a memory reduction of about half, and can fit into a single GPU memory for inference as shown in *Figure 9.4*:

```
from transformers import AutoModelForCausalLM

model = AutoModelForCausalLM.from_pretrained("EleutherAI/gpt-j-6B",
                                             torch_dtype=torch.float16,
                                             low_cpu_mem_usage=True)

model.num_parameters()

6050882784

param_size = 0
buffer_size = 0
for param in model.parameters():
    param_size += param.nelement() * param.element_size()

for buffer in model.buffers():
    buffer_size += buffer.nelement() * buffer.element_size()

size_all_mb = (param_size + buffer_size) / 1024**2
print('Size: {:.3f} GB'.format(size_all_mb/1000))
Size: 11.653 GB
```

Figure 9.4 – The model size and number of parameters for the GPT-J-6B model with FP16 precision

As shown in *Figure 9.3* and *Figure 9.4*, quantization can be a useful technique for reducing the memory footprint of the model.

Now, let's take a look at another technique, called model compilation.

Model compilation

Before we go into model compilation, let's first understand compilation.

Compilation is the process of converting a human-readable program that is a set of instructions into a machine-readable program. This core idea is used as a backbone for many programming languages, such as C, C++, and Java. This process also introduces efficiencies in the runtime environment of the program, such as making it platform-independent, reducing the memory size of the program, and so on. Most programming languages come with a **compiler**, which is used to compile the code into a machine-readable form as writing a compiler is a tedious process and requires a deep understanding of the programming language as well as the hardware.

A similar idea is used for compiling ML models. With ML, compilers place the core operations of the neural network on GPUs in a way that minimizes overhead. This reduces the memory footprint of the model, improves the performance efficiency, and makes it hardware agnostic. The concept is depicted in *Figure 9.5*, where the compiler takes a model developed in PyTorch, TensorFlow, XGBoost, and so on, converts it into an intermediate representation that is language agnostic, and then converts it into machine-readable code:

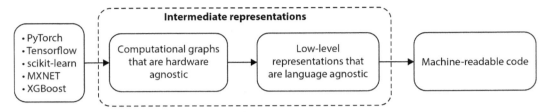

Figure 9.5 – High-level model compilation process

Model compilation eliminates the effort required to fine-tune the model for the specific hardware and software configurations of each platform. There are many frameworks available today using which you can compile your models, such as **TVM** and **ONNX**, each with its own pros and cons.

> **Note**
> Discussing compilers in detail is out of the scope of this book. For details on TVM, refer to this link: `https://tvm.apache.org/`. And for details about ONNX, refer to this link: `https://onnx.ai/`.

Let's discuss a feature of Amazon SageMaker called **SageMaker Neo**, which is used to optimize ML models for inference on multiple platforms. Neo automatically optimizes models written in various frameworks, such as Gluon, Keras, PyTorch, TensorFlow, and so on, for inference on different platforms, such as Linux and Windows, as well as many different processors. For a complete list of frameworks or processors supported by Neo, please refer to this link: `https://docs.aws.amazon.com/sagemaker/latest/dg/neo-supported-devices-edge.html`.

The way Neo works is that it will first read the model, convert the framework-specific operations and functions into an intermediate representation that is framework agnostic, and finally, apply a series of optimizations. It will then generate binary code for optimized operations, write it to the shared object library and save the model definition and parameters in separate files (`https://docs.aws.amazon.com/sagemaker/latest/dg/neo.html`).

It also provides a runtime for each target platform that loads and executes the compiled model. Moreover, it can optimize the models with parameters either in floating point 32 (FP32), quantized into integer 8 (INT8), or in floating point 16 (FP16). It can improve the model's performance up to 25 times with less than one-tenth of the footprint of a DL framework such as TensorFlow or PyTorch. To understand this further, let's take a pretrained image classification model from PyTorch and optimize it using SageMaker Neo.

> **Note**
>
> We touched on SageMaker Neo in *Chapter 8, Optimizing and Managing Machine Learning Models for Edge Deployment*. Here, we will be using the same example to explain it in detail. For the full code, refer to the GitHub link: `https://github.com/PacktPublishing/Applied-Machine-Learning-and-High-Performance-Computing-on-AWS/blob/main/Chapter08/sagemaker_notebook/1_compile_resnet_model_egde_manager.ipynb`.

Follow these steps to optimize a pretrained model using SageMaker Neo:

1. First, we will download a pretrained model from the PyTorch library as shown in the following code snippet:

```
...
from torchvision.models import resnet18, ResNet18_Weights
#initialize the model
weights = ResNet18_Weights.DEFAULT
model = resnet18(weights)
...
```

2. Next, we need to save the model in `model.tar.gz` format, which SageMaker requires, specify the input data shape, and upload the model to **Amazon S3**:

```
...
torch.save(model.state_dict(),
           './output/resnet18-model.pt')
with tarfile.open("model.tar.gz", "w:gz") as f:
    f.add("model.pth")
input_tensor = torch.zeros([1, 3, 224, 224])
model_uri = sagemaker_session.upload_data(
    path="model.tar.gz", key_prefix=key_prefix)
print("S3 Path for Model: ", model_uri)
...
```

3. Once we have the model in SageMaker format, we will prepare the parameters required for model compilation. Most importantly, you need to mention the `target_device` parameter, as based on it, SageMaker Neo will compile the model for the particular hardware on which the model will be deployed:

```
...
compilation_job_name = name_from_base("image-
classification-neo")
```

```
prefix = key_prefix+'/'+compilation_job_name+"/model"

data_shape = '{"input0":[1,3,224,224]}'
target_device = "ml_c5"
framework = "PYTORCH"
framework_version = "1.8"
compiled_model_path = "s3://{}/{}/output".format(bucket,
compilation_job_name)
print("S3 path for compiled model: ", compiled_model_
path)
...
```

4. Next, we will declare the PyTorchModel object provided by SageMaker, which will have the necessary configurations, such as the model's S3 path, the framework version, the inference script, the Python version, and so on:

```
...
from sagemaker.pytorch.model import PyTorchModel
from sagemaker.predictor import Predictor

sagemaker_model = PyTorchModel(
    model_data=model_uri,
    predictor_cls=Predictor,
    framework_version=framework_version,
    role=role,
    sagemaker_session=sagemaker_session,
    entry_point="inference.py",
    source_dir="code",
    py_version="py3",
    env={"MMS_DEFAULT_RESPONSE_TIMEOUT": "500"},
)
...
```

5. Finally, we will use the PyTorchModel object to create the compilation job and deploy the compiled model to the ml.c5.2xlarge instance, since the model was compiled for ml.c5 as the target device:

```
...
sagemaker_client = boto3.client("sagemaker",
```

```
        region_name=region)
target_arch = "X86_64"
target_os = 'LINUX'
response = sagemaker_client.create_compilation_job(
    CompilationJobName=compilation_job_name,
    RoleArn=role,
    InputConfig={
        "S3Uri": sagemaker_model.model_data,
        "DataInputConfig": data_shape,
        "Framework": framework,
    },
    OutputConfig={
        "S3OutputLocation": compiled_model_path,
        "TargetDevice": 'jetson_nano',
        "TargetPlatform": {
            "Arch": target_arch,
            "Os": target_os
        },
    },
    StoppingCondition={"MaxRuntimeInSeconds": 900},
)
print(response)
...
```

6. Once the model has finished compiling, you can then deploy the compiled model to make inference. In this case, we are deploying the model for **real-time inference** as an endpoint:

```
...
predictor = compiled_model.deploy(
    initial_instance_count=1,
    instance_type="ml.c5.2xlarge")
...
```

> **Note**
>
> For more details on different deployment options provided by SageMaker, refer to *Chapter 7, Deploying Machine Learning Models at Scale.*

7. Now, once the model is deployed, we can invoke the endpoint for inference as shown in the following code snippet:

```
...
import numpy as np
import json

with open("horse_cart.jpg", "rb") as f:
    payload = f.read()
    payload = bytearray(payload)

response = runtime.invoke_endpoint(
    EndpointName=ENDPOINT_NAME,
    ContentType='application/octet-stream',
    Body=payload,
    Accept = 'application/json')
result = response['Body'].read()
result = json.loads(result)
print(result)
...
```

> **Note**
> Since we have deployed the model as a real-time endpoint, you will be charged for the instance on which the model is deployed. Therefore, if you are not using the endpoint, make sure to delete it using the following code snippet.

8. Use the following code snippet to delete the endpoint, if you are not using it:

```
...
# delete endpoint after testing the inference
import boto3
# Create a low-level SageMaker service client.
sagemaker_client = boto3.client('sagemaker',
                                region_name=region)
# Delete endpoint
sagemaker_client.delete_endpoint(
    EndpointName=ENDPOINT_NAME)
...
```

Now that we understand how to optimize a model using SageMaker Neo for inference, let's talk about some of the key metrics that you should consider when trying to improve the latency of models. The ideas covered in the next section apply to DL models, even when you are not using SageMaker Neo, as you might not be able to compile all models with Neo. You can see the supported models and frameworks for SageMaker Neo here: `https://docs.aws.amazon.com/sagemaker/latest/dg/neo-supported-cloud.html`.

Key metrics for optimizing models

When it comes to real-time inference, optimizing a model for performance usually includes metrics such as latency, throughput, and model size. To optimize the model size, the process typically involves having a trained model, checking the size of the model, and if it does not fit into single CPU/GPU memory, you can choose any of the techniques discussed in the *Reducing the memory footprint of DL models* section to prepare it for deployment.

For deployment, one of the best practices is to standardize the environment. This will involve the use of containers to deploy the model, irrespective of whether you are deploying it on your own server or using Amazon SageMaker. The process is illustrated in *Figure 9.6*.

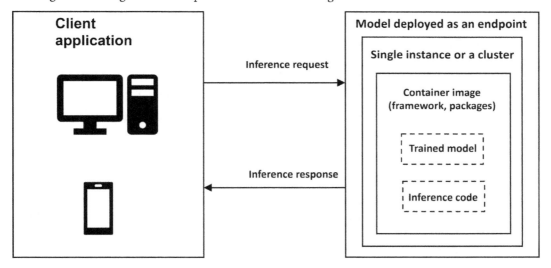

Figure 9.6 – Model deployed as a real-time endpoint

To summarize, we will first prepare the model for deployment, select or create a container to standardize the environment, followed by deploying the container on an instance(s). Therefore, to optimize the model's performance, it is important to look for both inference and instance metrics. Inference metrics include the following:

- **Invocations**: The number of requests sent to the model endpoint. You can get the total number of requests by using the sum statistics in **Amazon CloudWatch**, which monitors the AWS resources or applications that you run on AWS in real time, including SageMaker (https://docs.aws. amazon.com/AmazonCloudWatch/latest/monitoring/WhatIsCloudWatch. html).

- **Invocations per instance**: If your model is deployed on more than one machine, then it's important to understand the number of invocations sent to a model on each instance.

- **Model latency**: This is the time interval taken by a model to respond. It includes the local communication time taken to send the request, for the model to complete the inference in the container, and to get the response from the container of the model:

model latency = request time + inference time taken by the model + response time from the container

- **Overhead latency**: This is the time taken to respond after the endpoint has received the request minus the model latency. It can depend on multiple factors, such as request size, request frequency, authentication/authorization of the request, and response payload size.

- **Maximum Invocations**: This is the maximum number of requests to an endpoint per minute.

- **Cost per hour**: This gives the estimated cost per hour for your endpoint.

- **Cost per inference**: This provides the estimated cost per inference on your endpoint.

> **Note**
> Cost metrics provide the cost in US dollars.

In addition to inference metrics, you should also consider optimizing for instance metrics such as GPU utilization, GPU memory, CPU utilization, and CPU memory based on the instance type selected. Amazon SageMaker offers more than 70 instances from which you can choose to deploy your model. This brings up an additional question on how to determine which instance type and the number of instances to select for deploying the model in order to achieve your performance requirements. Let's discuss the approach for selecting the optimal instance for your model in the next section.

Choosing the instance type, load testing, and performance tuning for models

Traditionally, based on the model type (ML model or DL model) and model size, you can make a heuristic guess to test the model's performance on a few instances. This approach is fast but might not be the best approach. Therefore, in order to optimize this process, alternatively, you can use the **Inference Recommender** feature of Amazon SageMaker (https://docs.aws.amazon.com/sagemaker/latest/dg/inference-recommender.html), which automates the load testing and model tuning process across the SageMaker ML instances. It helps you to deploy the ML models on the optimized hardware, based on your performance requirements, at the lowest possible cost.

Let's take an example by using a pretrained image classification model to understand how Inference Recommender works. The following steps outline the process of using Inference Recommender:

1. Determine the ML model details, such as framework and domain. The following is a code snippet for this:

    ```
    ...
    # ML model details
    ml_domain = "COMPUTER_VISION"
    ml_task = "IMAGE_CLASSIFICATION"
    ...
    ```

2. Take a pretrained model and package it in the compressed TAR file (*.tar.gz) format, which SageMaker understands, and upload the model to Amazon S3. If you have trained the model on SageMaker, then you can skip this step:

    ```
    ...
    from torchvision.models import resnet18, ResNet18_Weights
    #initialize the model
    weights = ResNet18_Weights.DEFAULT
    model = resnet18(weights)
    torch.save(model.state_dict(), './output/resnet18-model.
    pt')
    with tarfile.open("model.tar.gz", "w:gz") as f:
        f.add("model.pth")
    input_tensor = torch.zeros([1, 3, 224, 224])
    model_uri = sagemaker_session.upload_data(path="model.
    tar.gz", key_prefix=key_prefix)
    print("S3 Path for Model: ", model_uri)
    ...
    ```

3. Select the inference container, which can be a prebuilt Docker container provided by AWS or your own custom container. In our example, we are fetching a prebuilt PyTorch container image provided by AWS:

```
...
instance_type = "ml.c5.xlarge"  # Note: you can use any
CPU-based instance type here, this is just to get a CPU
tagged image
container_uri = image_uris.retrieve(
    "pytorch",
    region,
    version=framework_version,
    py_version="py3",
    instance_type=instance_type,
    image_scope="inference",
)
container_uri
...
```

4. Create a sample payload. In our example, we have images in .jpg format, compress them in a TAR file, and upload it to Amazon S3. In this example, we are only using four images, but it's recommended to add a variety of samples, which is reflective of your actual payloads:

```
!wget https://multimedia-commons.s3-us-west-2.amazonaws.
com/data/images/139/019/1390196df443f2cf614f2255ae75fcf8.
jpg -P /sample-payload
!wget https://multimedia-commons.s3-us-west-2.amazonaws.
com/data/images/139/015/1390157d4caaf290962de5c5fb4c42.
jpg -P /sample-payload
!wget https://multimedia-commons.s3-us-west-2.amazonaws.
com/data/images/139/020/1390207be327f4c4df1259c7266473.
jpg  -P /sample-payload
!wget https://multimedia-commons.s3-us-west-2.amazonaws.
com/data/images/139/028/139028d865bafa3de66568eeb499f4a6.
jpg  -P /sample-payload
```

Compress the payload in TAR format as shown in the following code snippet and upload it to S3:

```
cd ./sample-payload/ && tar czvf ../{payload_archive_
name} *
```

Once, we have the payload in TAR format, let's use the following code snippet to upload it to Amazon S3:

```
...
sample_payload_url = sagemaker_session.upload_
data(path=payload_archive_name, key_prefix="tf_payload")
...
```

5. Register the model in the **model registry**, which is used to catalog models for production, manage model versions, associate metadata, manage the approval status of the model, deploy models to production, and automate the model deployment process (https://docs.aws.amazon.com/sagemaker/latest/dg/model-registry.html). Registering a model in the model registry is a two-step process, as shown here:

    I.   Create a model package group, which will have all the versions of the model:

```
...
model_package_group_input_dict = {
"ModelPackageGroupName": model_package_group_name,
"ModelPackageGroupDescription": model_package_group_
description,
}
create_model_package_group_response = sm_client.
create_model_package_group(**model_package_group_
input_dict)
...
```

    II. Register a model version to the model package group. To get the recommended instance type, you have two options – either you can specify a list of instances that you want Inference Recommender to use or you can not provide the instance list and it will pick the right instance based on the ML domain and ML task. For our example, we will use a list of common instance types used for image classification algorithms. This involves three steps.

    III. First, create an input dictionary with configuration for registering the model, as shown in the following code snippet:

```
...
model_approval_status = "PendingManualApproval"
# provide an input dictionary with configuration for
registering the model
model_package_input_dict = {
    "ModelPackageGroupName": model_package_group_name,
```

```
    "Domain": ml_domain.upper(),
    "Task": ml_task.upper(),
    "SamplePayloadUrl": sample_payload_url,
    "ModelPackageDescription": model_package_
description,
    "ModelApprovalStatus": model_approval_status,
}# optional - provide a list of instances
supported_realtime_inference_types = ["ml.c5.xlarge",
"ml.m5.large", "ml.inf1.xlarge"]

...
```

IV.   The second step is to create a model inference specification object, which will consist of providing details about the container, framework, model input, content type, and the S3 path of the trained model:

```
#create model inference specification object
modelpackage_inference_specification = {
    "InferenceSpecification": {
        "Containers": [
            {
                "Image": container_uri,
                "Framework": "PYTORCH",
                "FrameworkVersion": framework_version,
                "ModelInput": {"DataInputConfig":
data_input_configuration},
            }
        ],
        "SupportedContentTypes": "application/image",
        "SupportedRealtimeInferenceInstanceTypes":
supported_realtime_inference_types,  # optional
    }
}
# Specify the model data
modelpackage_inference_
specification["InferenceSpecification"]["Containers"]
[0]["ModelDataUrl"] = model_url
create_model_package_input_dict.update(modelpackage_
inference_specification)

...
```

V.   Finally, after providing the inference specification, we will then create the model package. Once the model package is created, you can then get the model package ARN, and it will also be visible in the SageMaker Studio UI, under **Model Registry**:

```
create_mode_package_response = sm_client.create_model_
package(**model_package_input_dict)
model_package_arn = create_mode_package_
response["ModelPackageArn"]
...
```

6.   Now, that the model has been registered, we will create an Inference Recommender job. There are two options – either you can create a default job to get instance recommendations or you can use an advanced job, where you can provide your inference requirements, tune environment variables, and perform more extensive load tests. An advanced job takes more time than a default job and depends on your traffic pattern and the number of instance types on which it will run the load tests.

In this example, we will create a default job, which will return a list of instance type recommendations including environment variables, cost, throughput, model latency, and the maximum number of invocations.

The following code snippet shows how you can create a default job:

```
...
response = sagemaker_client.create_inference_
recommendations_job(
    JobName=str(default_job),
    JobDescription="",
    JobType="Default",
    RoleArn=role,
    InputConfig={"ModelPackageVersionArn": model_package_
arn},
)
print(response)
...
```

> **Note**
>
> Code to create a custom load test is provided in the GitHub repository: `Applied-Machine-Learning-and-High-Performance-Computing-on-AWS/Chapter09/1_inference_recommender_custom_load_test.ipynb`.

In the next section, we will discuss the results provided by the Inference Recommender job.

Observing the results

The recommendation provided by Inference Recommender includes instance metrics, performance metrics, and cost metrics.

Instance metrics include InstanceType, InitialInstanceCount, and EnvironmentParameters, which are tuned according to the job for better performance.

Performance metrics include MaxInvocations and ModelLatency, whereas cost metrics include CostPerHour and CostPerInference.

These metrics enable you to make informed trade-offs between cost and performance. For example, if your business requirement is overall price performance with an emphasis on throughput, then you should focus on CostPerInference. If your requirement is a balance between latency and throughput, then you should focus on ModelLatency and MaxInvocations metrics.

You can view the results of the Inference Recommender job either through an API call or in the SageMaker Studio UI.

The following is the code snippet for observing the results:

```
...
data = [
    {**x["EndpointConfiguration"], **x["ModelConfiguration"],
**x["Metrics"]}
    for x in inference_recommender_
job["InferenceRecommendations"]
]
df = pd.DataFrame(data)
...
```

You can observe the results from the SageMaker Studio UI by logging into SageMaker Studio, clicking on the orange triangle icon, and selecting **Model registry** from the drop-down menu:

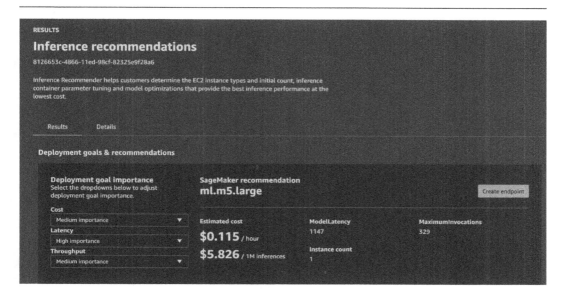

Figure 9.7 – Inference Recommender results

Now that we understand how the Inference Recommender feature of Amazon SageMaker can be used to get the right instance type and instance count, let's take a look at the topics covered in this chapter in the next section.

Summary

In this chapter, we discussed various techniques for optimizing ML and DL models for real-time inference. We talked about different ways to reduce the memory footprint of DL models, such as pruning and quantization, followed by a deeper dive into model compilation. We then discussed key metrics that can help in evaluating the performance of models. Finally, we did a deep dive into how you can select the right instance, run load tests, and automatically perform model tuning using SageMaker Inference Recommender's capability.

In the next chapter, we will discuss visualizing and exploring large amounts of data on AWS.

10
Data Visualization

In the previous chapters, we discussed various AWS services and tools that can help with building and running high-performance computation applications. We talked about the storage, compute instances, data processing, machine learning model building and hosting, and edge deployment of these applications. All these applications, especially those based around machine learning models, generally need some type of visualization. This visualization may vary from exploratory data analysis to model evaluation and comparison, to building dashboards showing various performance and business metrics.

Data visualization is very important for finding various business insights as well as deciding on what feature engineering steps to take to train a machine learning model that provides good results. AWS provides a few managed services to build data visualizations, as well as dashboards.

In this chapter, we are going to discuss one such option, Amazon SageMaker Data Wrangler, that enables users working in the domains of data science, machine learning, and analytics to build insightful data visualizations without writing much code. SageMaker Data Wrangler provides several in-built visualization options, along with the capability of adding custom visualizations with a few clicks and not much effort. This helps the data scientists with exploratory analysis, feature engineering, and experimentation processes involved in any data-driven use case.

In addition, we will also briefly touch upon the topic of AWS's graphics-optimized instances, since these instances can be used to create animated live data visualizations along with other high-performance computing applications such as game streaming and machine learning.

In this chapter, we'll cover the following topics:

- Data visualization using Amazon SageMaker Data Wrangler
- Amazon's graphics-optimized instances

Data visualization using Amazon SageMaker Data Wrangler

Amazon SageMaker Data Wrangler is a tool in SageMaker Studio that helps data scientists and machine learning practitioners carry out exploratory data analysis and feature engineering/transformation. SageMaker Data Wrangler is a low-code/no-code tool where users can either use built-in plotting or feature engineering capabilities or use code to make custom plots and carry out custom feature engineering. In data science projects with large datasets requiring visualization to carry out exploratory data analysis, SageMaker Data Wrangler can help build plots and visualizations very quickly with just a few clicks. We can import data from various data sources into Data Wrangler and also do operations such as joins and filtering. In addition, data insights and quality reports can also be generated to detect if there are any abnormalities in the data.

In this section, we will go through an example of how to build a workflow to carry out data analysis and visualization using SageMaker Data Wrangler.

SageMaker Data Wrangler visualization options

In SageMaker Data Wrangler, first, we need to import the data and then build a workflow to carry out various transformation and visualization tasks. At the time of writing, data can be imported into SageMaker Data Wrangler from Amazon S3, Amazon Athena, and Amazon Redshift. *Figure 10.1* shows the **Import data** screen in Amazon SageMaker Data Wrangler. To add data from Amazon Redshift, we need to click on the **Add data source** button near the top right:

Figure 10.1 – SageMaker Data Wrangler Data's import interface

Next, we will show an example of importing data into SageMaker Data Wrangler from Amazon S3, and then illustrate the various visualization options for this data. We are going to use the *Adult* dataset available at the University of California at Irvine's Machine Learning Repository (https://archive.ics.uci.edu/ml/datasets/Adult). This dataset classifies an individual into two classes: whether the person earns more than 50,000 dollars a year or not. There are various numerical and categorical features in the dataset. While importing the data, we can select whether we want to load the entire dataset or a sample of the dataset. *Figure 10.2* shows the workflow created by SageMaker Data Wrangler after importing the entire *Adult* dataset:

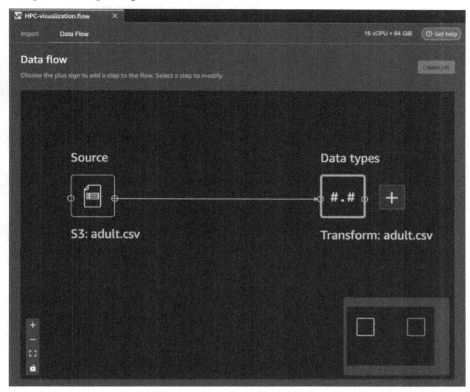

Figure 10.2 – Workflow created by SageMaker Data Wrangler after importing the data

By pressing the plus sign on the block on the right, we can add transformation and visualization steps to the workflow, as we are going to demonstrate next.

Adding visualizations to the data flow in SageMaker Data Wrangler

After importing the data, we can also view the data in SageMaker Data Wrangler, along with the column (variable) types. *Figure 10.3* shows a view of the Adult data table. This view can help data scientists and analysts quickly look at the data and validate it:

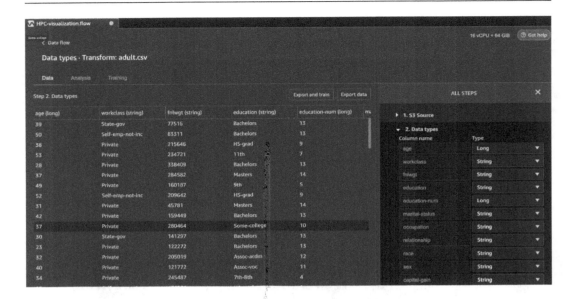

Figure 10.3 – Adult data table along with the column types

To construct various visualizations on our dataset, we must press the + sign on the **Transform: adult. csv** block shown in *Figure 10.2*. This shows us the various analysis and visualization options available in SageMaker Data Wrangler. We will now look at a few visualization examples while using SageMaker Data Wrangler on the *Adult* dataset.

Histogram

A **histogram** is one of the most commonly used plots that data scientists and analysts use in their exploratory data analysis, as well as for building the final reports of machine learning model results. SageMaker Data Wrangler provides the option of building histograms very quickly without writing any code. *Figure 10.4* shows a histogram plotted with SageMaker Data Wrangler, showing the distribution of the education-num (number of years of education) variable. We have also colored this distribution using the income target variable to show how income depends on the number of years of education:

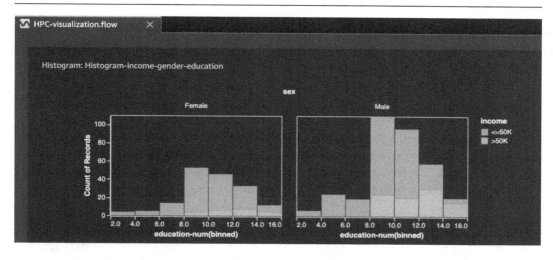

Figure 10.4 – Histogram of the number of education years (education-num), colored for the two classes (<=50k and > 50k)

In addition to color, we can also use the **Facet by** feature, which plots a histogram of one column for each value in another column. This is shown in *Figure 10.5*, where we have plotted the histogram of `education-num` colored by income. This is faceted by the `sex` variable to show the distribution for males and females separately:

Figure 10.5 – Histogram of the number of education years (education-num), colored for the two classes (<=50k and > 50k), and faceted by sex

We can see that creating histograms with multiple columns this way with just a few clicks and no need to write any code can be very useful for carrying out quick exploratory analysis and building reports in a short amount of time. This is often needed for several high-performance computation use cases, such as machine learning.

Scatter plot

With a **scatter plot**, we can plot the data points as a function of two or more variables using the X and Y axes, as well as color. Furthermore, like histograms, scatter plots can also be faceted by the values in an additional column. This type of data plot is useful when we want to see the relationship between variables and not just the distribution. *Figure 10.6* shows the scatter plot of age versus education-num, faceted by sex:

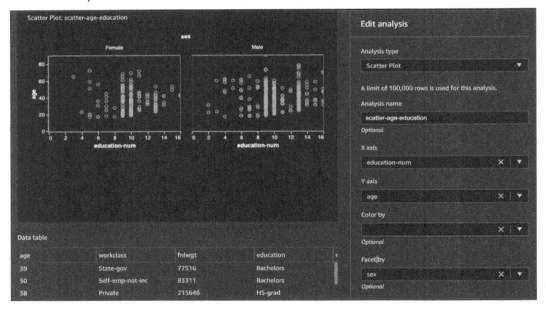

Figure 10.6 – Scatter plot of age versus education-num, faceted by sex, for the Adult dataset

Combined with histograms, scatter plots are extremely useful visualization tools that are used for exploratory data analysis in machine learning use cases.

Multicollinearity

In a dataset, **multicollinearity** arises when variables are related to each other. This is important in machine learning use cases since data dimensionality can be reduced if multicollinearity is detected. Data dimensionality reduction helps with avoiding the curse of dimensionality, as well as improving the performance of machine learning models. Furthermore, it also helps with reduced model training time, storage space, memory requirements, and data processing time during inference, hence cutting down on total cost. In SageMaker Data Wrangler, we can use the following methods to detect multicollinearity in our variables:

- **Variance Inflation Factor (VIF)**
- **Principal Component Analysis (PCA)**
- Lasso feature selection

Next, we will show examples of each of these methods using our *Adult* dataset in SageMaker Data Wrangler.

VIF

VIF indicates whether a variable is correlated to other variables or not. It is a positive number with a value of one indicating that the variable is uncorrelated to other variables in the dataset. A value greater than one means that the variable is correlated with other variables in the dataset. The higher the value, the higher the correlation with other variables. *Figure 10.7* shows the VIF plot for the numerical variables in the *Adult* dataset:

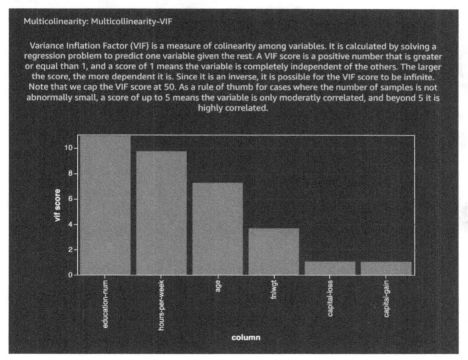

Figure 10.7 – VIF for the various numerical variables in the Adult dataset

As can be seen, education-num, hours-per-week, and age are highly correlated with other variables, whereas capital-gain and capital-loss do not correlate with other variables based on the VIF scores.

PCA

PCA is one of the most commonly used feature transformation and dimensionality reduction methods in machine learning. It is generally used not only as an exploratory data analysis tool but also as a preprocessing step in supervised machine learning problems. PCA projects data onto dimensions that are orthogonal to each other. The variables that are generated this way are ordered with decreasing variance (or singular values). These variances can be used to determine how much multicollinearity is in the variables. *Figure 10.8* shows the results of applying PCA to our *Adult* dataset using SageMaker Data Wrangler:

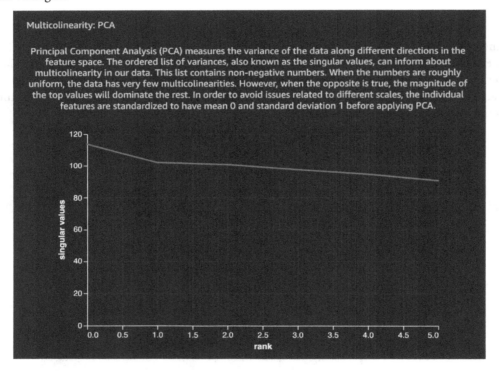

Figure 10.8 – Results of PCA on the Adult dataset using SageMaker Data Wrangler

As can be inferred from this figure, most of the variables do not have multicollinearity, whereas a few variables do.

Lasso feature selection

In SageMaker Data Wrangler, we can use lasso feature selection to find the most predictive variables for the target variable in our dataset. It uses the L1 regularization method to generate coefficients for each variable. A higher coefficient score means that the feature is more predictive of the target variable. Just like VIF and PCA, lasso feature selection is commonly used to reduce the dimensionality of datasets in machine learning use cases. *Figure 10.9* shows the results of applying lasso feature selection to our *Adult* dataset in SageMaker Data Wrangler:

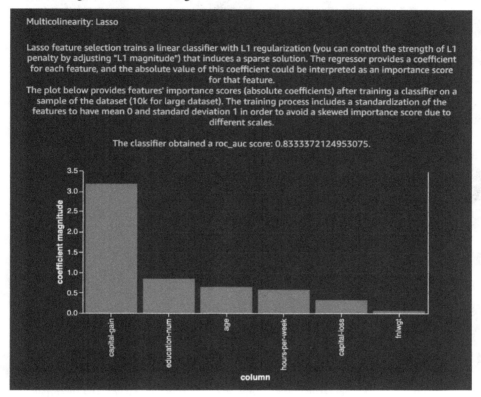

Figure 10.9 – Lasso feature selection results on the Adult dataset using SageMaker Data Wrangler

Next, we will discuss how we can also study variable importance using SageMaker Data Wrangler's Quick Model feature.

Quick Model

We can use **Quick Model** in Data Wrangler to evaluate variable/feature importance for our machine learning dataset. Quick Model trains a random forest regressor or random forest classifier, depending on the supervised learning problem type, and determines feature importance scores using the Gini importance method. The feature importance score is between 0 and 1, and a higher feature importance value indicates greater importance of that feature for the dataset. *Figure 10.10* shows the quick model plot created using SageMaker Data Wrangler for our *Adult* dataset:

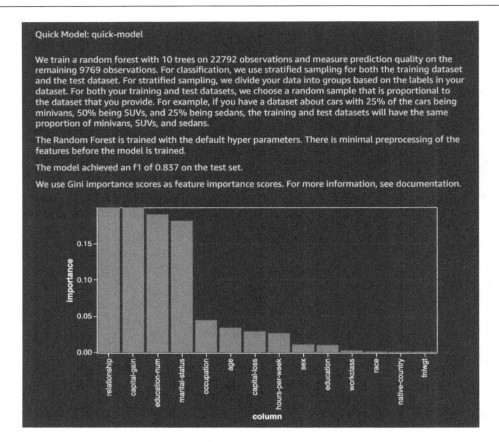

Figure 10.10 – Quick Model results for the Adult dataset using SageMaker Data Wrangler

Quick Model can help data scientists quickly evaluate the importance of features, and then use the results for dimensionality reduction, model performance improvement, or business insights.

Bias report

With a **bias report**, we can visualize if there is a potential bias or class imbalance in our dataset. This information can then be used to carry out class balancing, feature transformation, and model improvement. With SageMaker Data Wrangler, we can visualize class imbalance as well as several bias parameters for the dataset. *Figure 10.11* shows an example of a bias report for the sex variable (male or female), showing class imbalance and two other metrics:

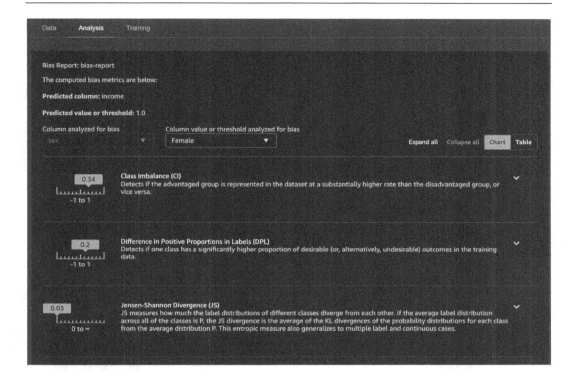

Figure 10.11 – Bias report for the Adult dataset using SageMaker Data Wrangler

For more information on these metrics, please refer to the *Further reading* section in this chapter.

Data quality and insights report

We can also build a data quality and insights report in SageMaker Data Wrangler. This report shows things such as a table summary, duplicate rows, the distribution of a target variable, anomalous samples, results of running a Quick Model, a confusion matrix (for classification problems), a feature summary along with feature importance, and feature detail plots. *Figure 10.12* shows the table summary for the *Adult* dataset:

SUMMARY

Dataset statistics

Key	Value		Feature type	Count
Number of features	15		numeric	6
Number of rows	32561		categorical	7
Missing	0%		text	0
Valid	100%		datetime	0
Duplicate rows	0.144%		binary	1
			unknown	0

Figure 10.12 – Table summary showing data statistics for the Adult dataset

Figure 10.13 shows the histogram of the target variable, along with various values of the occupation variable:

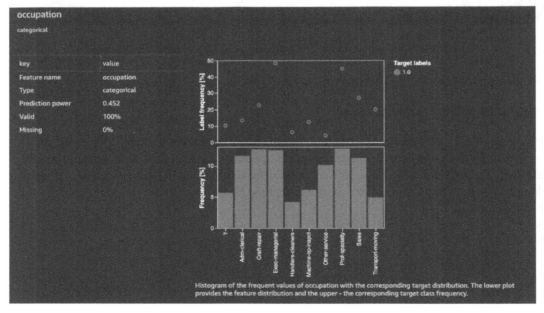

Figure 10.13 – Histogram of the target variable, along with various values of the occupation variable

Having access to these plots, metrics, and distributions with just a few clicks saves a lot of time and effort while performing exploratory data analysis.

Data flow

As we add steps to our data analysis in SageMaker Data Wrangler, they show up in the data flow. The data flow visualization shown in *Figure 10.14* is for the group of transformations that we have carried out in this flow so far:

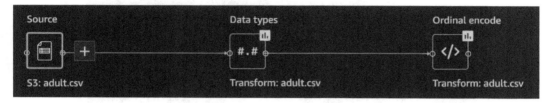

Figure 10.14 – SageMaker Data Wrangler flow showing the groups
of steps we have performed on our dataset

We can click on the individual boxes to see the exploratory analysis steps and data transformations carried out so far in the workflow. For example, clicking on the center box shows us the steps shown in *Figure 10.15*:

Figure 10.15 – Various exploratory data analysis and data transformation
steps carried out on our dataset in SageMaker Data Wrangler

Clicking on the right-hand box shown in *Figure 10.14* shows us our bias and data quality reports, as shown in *Figure 10.16*:

Figure 10.16 – Bias report and data quality and insights report steps in a SageMaker Data Wrangler flow

SageMaker Data Wrangler allows the results of these steps to be exported to the SageMaker Feature Store or Amazon S3. Furthermore, we can also export these steps as code to SageMaker pipelines or as Python code.

In this section, we discussed the various visualization options in Amazon SageMaker using SageMaker Data Wrangler. There is a variety of built-in visualization and exploratory analysis options in Data Wrangler that we can use in our machine learning use cases. In the next section, we are going to discuss Amazon's graphics-optimized instance options.

Amazon's graphics-optimized instances

Amazon has a variety of graphics-optimized instances that can be used in high-performance computing applications such as machine learning use cases and applications that require graphics-intensive computation workload. These instances are available with either NVIDIA GPUs or AMD GPUs and are required for high-performance computation use cases such as game streaming, graphics rendering, machine learning, and so on.

Benefits and key features of Amazon's graphics-optimized instances

In this section, we will outline a few of the benefits and key features of Amazon's graphics-optimized instances:

- **High performance and low cost**: Machines with good GPUs are generally quite expensive to purchase and difficult to scale due to their high cost. Amazon provides options to get high-performance instances equipped with state-of-the-art GPUs at low cost. These instances can be used to run graphics-intensive applications, build visualizations, and carry out machine learning training and inference. Also, these instances provide over a terabyte of NVMe-based solid-state storage for very fast access to local data, which is often needed in high-performance computation use cases.

  While AWS has a very wide variety of GPU-based instances that can be used for different types of high-performance computing applications, P3, P3dn, P4, and G4dn instances are especially suited for carrying out distributed machine learning model training tasks on multiple nodes. In addition, these instances can provide up to 400 gigabits per second of network bandwidth for applications that have very high throughput requirements.

- **Fully managed offerings**: Amazon's graphics-optimized instances can be provisioned as **Elastic Compute Cloud** (**EC2**) instances for a variety of use cases, including machine learning, numerical optimization, graphics rendering, gaming, and streaming. Users can install custom kernels and libraries and manage them as needed. These instances also support Amazon machine images for common deep learning frameworks such as TensorFlow, PyTorch, and MXNet. In addition, users can use these instances within Amazon SageMaker to train deep learning models. When used within SageMaker for training jobs, these instances are fully managed by AWS and are used only for the duration of the training job, thus reducing the cost significantly.

In the following section, we will summarize what we have learned in this chapter.

Summary

In this chapter, we discussed how to build quick visualizations for analytics and machine learning use cases using Amazon SageMaker Data Wrangler. We showed a few of the various exploratory data analysis, plotting, and data transformation options available within SageMaker Data Wrangler. The ability to quickly and easily build these visualizations and bias and quality reports is very important for data scientists and practitioners in the machine learning domain since it helps in cutting down on the cost and effort associated with exploratory data analysis significantly. In addition, we discussed Amazon's graphics-optimized instances that are available for high-performance computing applications such as game streaming, rendering, and machine learning use cases.

From the next chapter onwards, we will start discussing various applications of high-performance computing and applied machine learning, with the first one being computational fluid dynamics.

Further reading

To learn more about the topics we discussed in this chapter, please refer to the following resources:

- Adult dataset at UC Irvine's Machine Learning Repository: `https://archive.ics.uci.edu/ml/datasets/Adult`

- Amazon SageMaker Data Wrangler: `https://aws.amazon.com/sagemaker/data-wrangler/`

- Amazon SageMaker Data Wrangler documentation: `https://docs.aws.amazon.com/sagemaker/latest/dg/data-wrangler.html`

- Amazon SageMaker Data Wrangler visualization: `https://docs.aws.amazon.com/sagemaker/latest/dg/data-wrangler-analyses.html`

- Random forest regressor used in SageMaker Data Wrangler: `https://spark.apache.org/docs/latest/ml-classification-regression.html#random-forest-regression`

- Random forest classifier used in SageMaker Data Wrangler: `https://spark.apache.org/docs/latest/ml-classification-regression.html#random-forest-classifier`

- Data bias reports: `https://docs.aws.amazon.com/sagemaker/latest/dg/data-bias-reports.html`

- Amazon EC2 G4 instances: `https://aws.amazon.com/ec2/instance-types/g4/`

Part 3:
Driving Innovation
Across Industries

AWS provides high-performance storage, compute, and networking services to serve typical **high-performance computing** (HPC) applications that span various verticals, including multi-physics modeling, genomics, computer-aided design and simulation, and weather modeling and prediction, as well as applications such as **machine learning** (ML). In this part, we cover four major HPC applications, how they are set up and solved on AWS, and end each chapter with how emerging research in ML is used in conjunction with the classic ways to solve these typical HPC problems.

This part comprises the following chapters:

- *Chapter 11, Computational Fluid Dynamics*
- *Chapter 12, Genomics*
- *Chapter 13, Autonomous Vehicles*
- *Chapter 14, Numerical Optimization*

11
Computational Fluid Dynamics

Computational Fluid Dynamics (**CFD**) is a technique used to analyze how fluid (air, water, and other fluids) flow over or inside objects of interest. CFD is a mature field that originated several decades ago and is used in fields of study related to manufacturing, healthcare, the environment, and aerospace and automotive industries that involve fluid flow, chemical reactions, or thermodynamic reactions and simulations. Given the field's progress and long history, it is beyond the scope of this book to discuss many aspects of this field. However, these video links may be a great way for readers to get up to speed on what CFD is and some CFD tools, and best practices on AWS:

- `https://www.youtube.com/watch?v=__7_aHrNUF4&ab_channel=AWSPublicSector`

- `https://www.youtube.com/watch?v=8rAvNbCJ7M0&ab_channel=AWSOnlineTechTalks`

In this chapter, we are going to review the field of CFD and provide insights into how **Machine Learning** (**ML**) is being used today with CFD. Additionally, we will examine some of the ways you can run CFD tools on AWS.

We will cover the following topics in this chapter:

- Introducing CFD

- Reviewing best practices for running CFD on AWS

- Discussing how ML can be applied to CFD

Technical requirements

You should have the following prerequisites before getting started with this chapter:

- Familiarity with AWS and its basic usage.

- A web browser (for the best experience, it is recommended that you use a Chrome or Firefox browser).

- An AWS account (if you are unfamiliar with how to get started with an AWS account, you can go to this link: `https://aws.amazon.com/getting-started/`).

- Some familiarity with CFD. Although we will provide a brief overview of CFD, this chapter is best suited for readers that are at least aware of some of the typical use cases that can be solved using CFD.

In the following section, we will introduce CFD through an example application problem – designing a race car!

Introducing CFD

CFD is the prediction of fluid flow using numerical analysis. Let's break that down:

- **Prediction**: Just as with other physical phenomena, fluid flow can be modeled mathematically, and simulated. For readers from the field of ML, this is different from an ML model's *prediction*. Here, we solve a set of equations iteratively to construct the flow inside or around a body. Mainly, we use **Navier-Stokes** equations.

- **Numerical analysis**: Several tools have been created to help actually solve these equations – not surprisingly, these tools are called **solvers**. As with any set of tools, there are commercial and **open source** varieties of these solvers. It is uncommon nowadays to write any code related to the actual solving of equations – similar to how you don't write your own ML framework before you start solving your ML problems. Numerical or mathematical methods that have been studied for decades are implemented through code in these solvers that help with the analysis of fluid flow.

Now, imagine you are the team principal for a new **Formula 1 (F1)** team who is responsible for managing the design of a new car for the upcoming race season. The design of this car has to satisfy many new F1 regulations that define constraints of how the car can be designed. Fortunately, you have a large engineering team that can manage the design and manufacturing of a new car that is proposed. The largest teams spend millions of dollars on just the conceptual design of the car before even manufacturing a single part. It is typical for teams to start with a baseline design and improve this design iteratively. This iterative improvement of design is not unique to racing car development; think of the latest version of the iPhone in your pocket or purse, or how generations of commercial passenger aircraft designs look similar but are very different. You task your engineers with designing modifications to the existing car using **Computer-Aided Design (CAD)** tools, and after a month of working on potential design changes, they show you the design of your team's latest car (see *Figure 11.1*). This looks great!

Figure 11.1 – F1 car design

However, how do you know whether this car will perform better on track? Two key metrics that you can track are as follows:

- **Drag**: The resistance caused by an object in fluid flow. The **coefficient of drag** is a dimensionless quantity that is used to quantify drag. For your F1 car, a higher coefficient of drag is worse since your car will move slower, considering all the other factors remain constant.

- **Downforce**: Aerodynamic forces that push the car onto the track; the higher the downforce, the better it is since it provides greater grip when traveling or turning at high speeds.

Figure 11.2 shows the direction of these two forces applied to the F1 car:

Figure 11.2 – Drag and downforce directions on the F1 car

Now, one way to measure drag and downforce is to manufacture the entire car, drive around a track with force sensors, and report this back to the team – but what if you had a different design in mind? Or a variation in one of the components of your car? You would have rebuilt these components, or the entire car, and then perform the same tests, or run a scale model in a wind tunnel – these options can be very time-consuming and very expensive. This is where numerical analysis codes such as CFD tools become useful. With CFD tools, you can simulate different flow conditions over the car and calculate the drag and downforce.

It is typical in CFD to create a **flow domain** with the object of interest inside it. This can look similar to *Figure 11.3* for **external flows** (for example, the flow around the vehicle). On the other hand, you could have **internal flows** where the domain is defined in the object itself (such as the flow inside a bent pipe). In *Figure 11.3*, the green and blue surfaces represent the **inlet** and **outlet** in this domain. Air flows from the **inlet**, over and around the car, and out through the **outlet**.

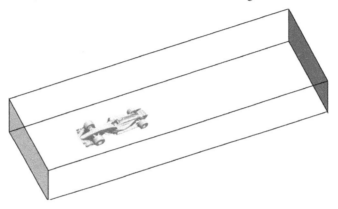

Figure 11.3 – CFD domain defined around the F1 car

The car and the domain so far are conceptual ideas and need to be represented as objects or files that CFD code can ingest and use. A typical file format used to represent objects is the **Stereolithography** (**STL**) file format. Each object is represented as a set of triangles, and each triangle is represented by a set of 3D points. The same car in an STL format is shown in *Figure 11.4* – the car is now a collection of tens of thousands of triangles.

Figure 11.4 – F1 car in an STL format

We can now use this car object and **mesh** the CFD domain. Creating a **mesh**, or **meshing**, is the process of creating grid points in the CFD domain where numerical equations related to fluid flow are to be solved. Meshing is a very important process, as this can directly influence results, and also sometimes cause the numerical simulation to diverge or not solve.

> **Note**
>
> Meshing techniques and details about the algorithms used are beyond the scope of this book. Each solver tool implements different meshing techniques with various configurations. Teams spend a significant amount of time getting high-quality meshes while balancing the complexity of the mesh to ensure faster solving times.

Once the mesh is built out, it may look similar to *Figure 11.5*. We see that there is a concentration of grid cells closer to the body. Note that this is a slice of the mesh, and the actual mesh is a 3D volume with the bounds that were defined in *Figure 11.3*.

Figure 11.5 – CFD mesh built for the F1 car case

Once the mesh is built out, we can use CFD solvers to calculate the flow around this F1 car, and then post-process these results to provide us with predictions for drag and downforce. *Figure 11.6* and *Figure 11.7* show typical post-processed images involving streamlines (the white lines in the images representing how fluid flows around the body), a velocity slice (the magnitude of the velocity on a plane or cross-section of interest), pressure on the car body (redder regions are higher pressure), and the original car geometry for context.

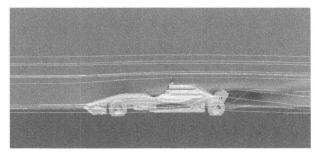

Figure 11.6 – Post-processed results for the F1 car case showing streamlines and velocity slice

Figure 11.7 shows a different output visualization of pressure on the surface of the car, along with the streamlines in a perspective view.

Figure 11.7 – Post-processed results for the F1 car case showing pressure on the car body with streamlines

In summary, running a CFD case involves the following:

1. Loading and manipulating the geometry

2. Meshing the CFD domain

3. Using a solver to solve for the flow in the domain

4. Using post-processing tools to visualize the results

In the next section, we will discuss a few ways of running CFD analyses on AWS according to our documented best practices.

Reviewing best practices for running CFD on AWS

CFD, being very compute-intensive, needs to be scaled massively to be practical for companies that depend on analysis results to make decisions about their product designs. AWS allows customers to run CFD simulations at a massive scale (thousands of cores), on-demand, with multiple commercial and **open source** tools, and without any capacity planning or up-front capital investment. You can find many useful links related to CFD on AWS here: https://aws.amazon.com/hpc/cfd/.

As highlighted at the outset of this chapter, there are several commercial and open source tools available to solve your CFD problems that run at scale on AWS. Some of these tools are as follows:

*   Siemens SimCenter STAR-CCM+

*   Ansys Fluent

*   OpenFOAM (open source)

In this chapter, we will be providing you with examples of how to set up and use *OpenFOAM*. For other tools, please take a look at this workshop provided by AWS: `https://cfd-on-pcluster.workshop.aws/`.

> **Note**
>
> Note that the AWS **Well-Architected** pillar defines best practices for running any kind of workload on AWS. It includes the best practices for designing architectures on AWS with the following pillars: **Operational Excellence**, **Security**, **Reliability**, **Performance Efficiency**, **Cost Optimization**, and **Sustainability**.
>
> If you are unfamiliar with the Well-Architected Framework, you can read about it in detail here: `https://docs.aws.amazon.com/wellarchitected/latest/framework/welcome.html/`.

We will now discuss two different ways of running CFD simulations on AWS: using ParallelCluster and using CFD Direct.

Using AWS ParallelCluster

On AWS, these Well-Architected best practices are encapsulated in a solution called AWS ParallelCluster that you can launch in your AWS account. ParallelCluster lets you configure and launch an entire HPC cluster with a simple **Command-Line Interface (CLI)**. The CLI also allows you to dynamically scale resources needed for your CFD (and other HPC) applications as needed, in a secure manner. Popular schedulers such as **AWS Batch** or **Slurm** can be used to submit and monitor jobs on ParallelCluster. Here are some steps to follow for installing ParallelCluster (note that a complete set of steps can be found on the official AWS documentation page for ParallelCluster here: `https://docs.aws.amazon.com/parallelcluster/latest/ug/install-v3-pip.html`).

Step 1 – creating an AWS Cloud9 IDE

This helps us have access to a full IDE on the cloud on a specified instance type, with temporary, managed credentials that can be used to launch ParallelCluster. Follow the instructions here to launch an AWS Cloud9 IDE: `https://docs.aws.amazon.com/cloud9/latest/user-guide/setup-express.html`.

Once you have created your Cloud9 IDE, navigate to the terminal as shown in the instructions here: `https://docs.aws.amazon.com/cloud9/latest/user-guide/tour-ide.html#tour-ide-terminal`.

Step 2 – installing the ParallelCluster CLI

Once you are inside the terminal, do the following:

1. Use `pip` to install `ParallelCluster`:

    ```
    . . .
    python3 -m pip install "aws-parallelcluster" --upgrade
    --user
    . . .
    ```

2. Next, make sure that you have **Node Version Manager** (**NVM**) installed:

    ```
    . . .
    curl -o- https://raw.githubusercontent.com/nvm-sh/nvm/
    v0.38.0/install.sh | bash
    chmod ug+x ~/.nvm/nvm.sh
    source ~/.nvm/nvm.sh
    nvm install --lts
    node -version
    . . .
    ```

3. Lastly, verify that `ParallelCluster` has been installed successfully:

    ```
    pcluster version

    {
        "version": "3.1.3"
    }
    ```

Let's move on to *step 3*.

Step 3 – configuring your ParallelCluster

Before you launch ParallelCluster, you need to define parameters using the `configure` command:

```
. . .
pcluster configure
. . .
```

The command-line tool will ask you the following questions for creating a configuration (or config, for short) file:

- Region in which to set up ParallelCluster (for example, US-East-1)
- **EC2 Key Pair** to use (learn more about **Key Pairs** here: https://docs.aws.amazon. com/AWSEC2/latest/UserGuide/ec2-key-pairs.html
- Operating system (for example, Amazon Linux 2, CentOS 7, or Ubuntu)
- Head node instance type (for example, **c5n.18xlarge**)
- Whether to automate VPC creation
- Subnet configuration (for example, head or main node placed in a public subnet with the rest of the compute fleet in a private subnet or subnets)
- Additional shared storage volume (for example, FSx configuration)

This creates a config file that can be found in ~/.parallelcluster and modified before the creation of the cluster. Here is an example of a ParallelCluster config file:

```
...
Region: us-east-2
Image:
  Os: alinux2
HeadNode:
  InstanceType: c5.4xlarge
  Networking:
    SubnetId: subnet-0e6e79abb7ed2452c
  Ssh:
    KeyName: pcluster
Scheduling:
  Scheduler: slurm
  SlurmQueues:
  - Name: queue1
    ComputeResources:
    - Name: c54xlarge
      InstanceType: c5.4xlarge
      MinCount: 0
      MaxCount: 4
    - Name: m516xlarge
      InstanceType: m5.16xlarge
```

```
        MinCount: 0
        MaxCount: 2
      Networking:
        SubnetIds:
        - subnet-09299f6d9ecfb8122
  ...
```

A deeper dive into the intricacies of the ParallelCluster config file can be found here: `https://aws.amazon.com/blogs/hpc/deep-dive-into-the-aws-parallelcluster-3-configuration-file/`.

Step 4 – launching your ParallelCluster

Once you have verified the config file, use the following command to create and launch `ParallelCluster`:

```
...
pcluster create-cluster --cluster-name mycluster --cluster-
configuration config
{
  "cluster": {
    "clusterName": "mycluster",
    "cloudformationStackStatus": "CREATE_IN_PROGRESS",
    "cloudformationStackArn": "arn:aws:cloudformation:us-
east-2:989279443319:stack/mycluster/c6fdb600-d49e-11ec-9c26-
069b96033f9a",
    "region": "us-east-2",
    "version": "3.1.3",
    "clusterStatus": "CREATE_IN_PROGRESS"
  }
}...
```

Here, our cluster has been named `mycluster`. This will launch a CloudFormation template with the required resources to work with ParallelCluster, based on the config file you previously defined. The following services are used by AWS ParallelCluster:

- AWS Batch
- AWS CloudFormation
- Amazon CloudWatch
- Amazon CloudWatch Logs

- AWS CodeBuild

- Amazon DynamoDB

- Amazon Elastic Block Store

- Amazon **Elastic Compute Cloud (EC2)**

- Amazon Elastic Container Registry

- Amazon **Elastic File System (EFS)**

- Amazon FSx for Lustre

- AWS Identity and Access Management

- AWS Lambda

- NICE DCV

- Amazon Route 53

- Amazon Simple Storage Service

- Amazon VPC

For more details on the services used, please refer to the links provided in the *References* section of this chapter. A simplified architecture diagram of AWS ParallelCluster is shown in *Figure 11.8* – more details can be found on the following blog: `https://aws.amazon.com/blogs/compute/running-simcenter-star-ccm-on-aws/`. Otherwise, see the documentation page for ParallelCluster (`https://docs.aws.amazon.com/parallelcluster/latest/ug/what-is-aws-parallelcluster.html`).

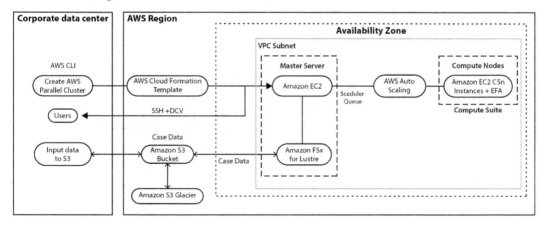

Figure 11.8 – AWS ParallelCluster architecture

The launch will typically take about 10 minutes and can be tracked both on the console as well as on the CloudFormation page on the AWS Management Console. On the console, the following message will confirm that your launch is in progress:

```
pcluster list-clusters --query
'clusters[?clusterName==`mycluster`]'
[
  {
    "clusterName": "mycluster",
    "cloudformationStackStatus": "CREATE_IN_PROGRESS",
    "cloudformationStackArn": "arn:aws:cloudformation:us-east-
2:<account_number>:stack/mycluster/c6fdb600-d49e-11ec-9c26-
069b96033f9a",
    "region": "us-east-2",
    "version": "3.1.3",
    "clusterStatus": "CREATE_IN_PROGRESS"
  }
]
```

Wait for the status to say `"clusterStatus": "CREATE_COMPLETE"`

Step 5 – installing OpenFOAM on the cluster

To install OpenFOAM on your cluster, see the following:

1. First, add **Secure Shell (SSH)** into the head node of your newly created ParallelCluster:

    ```
    pcluster ssh -n mycluster -i pcluster.pem

    The authenticity of host '3.135.195.149 (3.135.195.149)'
    can't be established.
    ECDSA key fingerprint is
    SHA256:DZPeIcVRpZDg3VMYhA+2zAvEoLnD3gI6mLVkMPkyg90.
    ECDSA key fingerprint is
    MD5:87:16:df:e7:26:f5:a0:da:a8:3a:7c:c4:c8:92:60:34.
    Are you sure you want to continue connecting (yes/no)?
    yes

    Warning: Permanently added '3.135.195.149' (ECDSA) to the
    list of known hosts.
    Last login: Sun May 15 22:36:41 2022
    ```

```
   __|   __|_  )
   _|   (    /    Amazon Linux 2 AMI
  ___|\___|___|
```

2. You are now in the head node of the ParallelCluster. Next, download the OpenFOAM files as follows:

   ```
   . . .
   wget https://sourceforge.net/projects/openfoam/files/
   v2012/OpenFOAM-v2012.tgz
   wget https://sourceforge.net/projects/openfoam/files/
   v2012/ThirdParty-v2012.tgz
   . . .
   ```

3. Next, untar the two files you just downloaded:

   ```
   . . .
   tar -xf OpenFOAM-v2012.tgz
   tar -xf ThirdParty-v2012.tgz
   . . .
   ```

4. Change the directory to the newly extracted OpenFOAM folder and compile OpenFOAM:

   ```
   . . .
   cd OpenFOAM-v2012
   export WM_NCOMPPROCS=36
   ./Allwmake
   . . .
   ```

To install OpenFOAM on all nodes, you can use the `sbatch` command, and submit the preceding commands as a file named `compile.sh`: for example, `sbatch compile.sh`.

Once installation completes, you can run a sample CFD application as shown in *step 6*.

Step 6 – running a sample CFD application

Here, we will run a sample CFD application using ParallelCluster. First, we access the head node of the cluster we just created using SSH:

```
. . .
pcluster ssh --cluster-name mycluster -i /path/to/keyfile.pem
. . .
```

Make sure you use the same `.pem` file that you created in *step 3*!

In this case, we will be running an example from OpenFOAM – incompressible flow over a motorbike. The case files for this case can be found here: `https://static.us-east-1.prod.workshops. aws/public/a536ee90-eecd-4851-9b43-e7977e3a5929/static/motorBikeDemo. tgz`.

The geometry corresponding to this case is shown in *Figure 11.9*.

Figure 11.9 – Geometry for the motorbike case in OpenFOAM

To run the case in just the head node, you can run the following commands:

```
cp $FOAM_TUTORIALS/resources/geometry/motorBike.obj.gz
constant/triSurface/
surfaceFeatureExtract
blockMesh
snappyHexMesh
checkMesh
potentialFoam
simpleFoam
```

We will go into detail about what these commands do in later sections. For now, our aim is just to run the example motorbike case.

To run the same case in parallel with all your compute nodes, you can use sbatch to submit the following shell script (similar to submitting the installation shell script). We can define some input arguments to the script, followed by loading **OpenMPI** and **OpenFOAM**:

```
...
#!/bin/bash
#SBATCH --job-name=foam
#SBATCH --ntasks=108
#SBATCH --output=%x_%j.out
#SBATCH --partition=compute
#SBATCH --constraint=c5.4xlarge

module load openmpi
source OpenFOAM-v2012/etc/bashrc


cp $FOAM_TUTORIALS/resources/geometry/motorBike.obj.gz
constant/triSurface/
```

First, we mesh the geometry using the blockMesh and snappyHexMesh tools (see the following code):

```
surfaceFeatureExtract  > ./log/surfaceFeatureExtract.log 2>&1
blockMesh  > ./log/blockMesh.log 2>&1
decomposePar -decomposeParDict system/decomposeParDict.
hierarchical  > ./log/decomposePar.log 2>&1
mpirun -np $SLURM_NTASKS snappyHexMesh -parallel -overwrite
-decomposeParDict system/decomposeParDict.hierarchical  > ./
log/snappyHexMesh.log 2>&1
```

We then check the quality of the mesh using checkMesh, and renumber and print out a summary of the mesh (see code):

```
mpirun -np $SLURM_NTASKS checkMesh -parallel -allGeometry
-constant -allTopology -decomposeParDict system/
decomposeParDict.hierarchical > ./log/checkMesh.log 2>&1
mpirun -np $SLURM_NTASKS redistributePar -parallel -overwrite
-decomposeParDict system/decomposeParDict.ptscotch > ./log/
decomposePar2.log 2>&1
mpirun -np $SLURM_NTASKS renumberMesh -parallel -overwrite
```

```
-constant -decomposeParDict system/decomposeParDict.ptscotch >
./log/renumberMesh.log 2>&1
mpirun -np $SLURM_NTASKS patchSummary -parallel
-decomposeParDict system/decomposeParDict.ptscotch > ./log/
patchSummary.log 2>&1
ls -d processor* | xargs -i rm -rf ./{}/0
ls -d processor* | xargs -i cp -r 0.orig ./{}/0
```

Finally, we run OpenFOAM through the potentialFoam and simpleFoam binaries as shown here:

```
mpirun -np $SLURM_NTASKS potentialFoam -parallel
-noFunctionObjects -initialiseUBCs -decomposeParDict system/
decomposeParDict.ptscotch > ./log/potentialFoam.log 2>&1s
mpirun -np $SLURM_NTASKS simpleFoam
-parallel  -decomposeParDict system/decomposeParDict.ptscotch >
./log/simpleFoam.log 2>&1
...
```

You can follow instructions in the following AWS workshop to visualize results from the CFD case: https://catalog.us-east-1.prod.workshops.aws/workshops/21c996a7-8ec9-42a5-9fd6-00949d151bc2/en-US/openfoam/openfoam-visualization.

Let's discuss CFD Direct next.

Using CFD Direct

In the previous section, we saw how you can run a CFD simulation using ParallelCluster on AWS. Now, we will look at how to run CFD using the CFD Direct offering on AWS Marketplace: https://aws.amazon.com/marketplace/pp/prodview-ojxm4wfrodtj4. CFD Direct provides an Amazon EC2 image built on top of Ubuntu with all the typical tools you need to run CFD with OpenFOAM.

Perform the following steps to get started:

1. Follow the link above to CFD Direct's Marketplace offering, and click **Continue to Subscribe**.

2. Then, follow the instructions provided and click **Continue to Configure** (leave all options as the default), and then **Continue to Launch**. Similar to ParallelCluster, remember to use the right EC2 key pair so you can SSH into the instance that is launched for you.

Figure 11.10 – CFD Direct AWS Marketplace offering (screenshot taken as of August 5, 2022)

Follow the instructions and get more help on using CFD Direct's image here: `https://cfd.direct/cloud/aws/`.

To connect to the instance for the first time, use the instructions given here: `https://cfd.direct/cloud/aws/connect/`.

In the following tutorial, we will use the NICE DCV client as a remote desktop to interact with the EC2 instance.

To install NICE DCV, perform the following steps:

1. First, SSH into the instance you just launched, and then download and install the server. For example, with Ubuntu 20.04, use the following command:

    ```
    wget https://d1uj6qtbmh3dt5.cloudfront.net/nice-dcv-ubuntu2004-x86_64.tgz
    ```

2. Then, execute the following command to extract the `tar` file:

    ```
    tar -xvzf nice-dcv-2022.0-12123-ubuntu2004-x86_64.tgz && cd nice-dcv-2022.0-12123-ubuntu2004-x86_64
    ```

3. Install NICE DCV by executing the following:

    ```
    sudo apt install ./nice-dcv-server_2022.0.12123-1_amd64.ubuntu2004.deb
    ```

4. To start the NICE DCV server, use the following command:

    ```
    sudo systemctl start dcvserver
    ```

5. Finally, start a session using the following:

    ```
    dcv create-session cfd
    ```

6. Find the public IP of your launched EC2 instance and use any NICE DCV client to connect to the instance (see *Figure 11.11*):

Connection has been closed: network error

Connect

Connection Settings

Figure 11.11 – Connecting to the EC2 instance using a public IP

7. Next, use the username and password for Ubuntu (see *Figure 11.12*). If you haven't set a password, use the `passwd` command on a terminal using SSH.

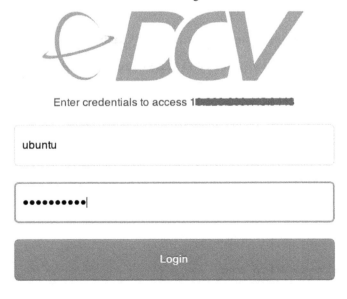

Enter credentials to access 1█████████████████

ubuntu

•••••••••|

Login

Cancel

Figure 11.12 – Entering a username and password for Ubuntu

8. If prompted, select the session you want to connect to. Here, we started a session called cfd. You should now be looking at your Ubuntu desktop with OpenFOAM 9 preinstalled.

Figure 11.13 – Ubuntu desktop provided by CFD Direct

9. To locate all the OpenFOAM tutorials to try out, use the following command:

```
echo $FOAM_TUTORIALS
/opt/openfoam9/tutorials
```

10. We will run a basic airfoil tutorial that is located in the following directory:

```
/opt/openfoam9/tutorials/incompressible/simpleFoam/
airFoil2D/
```

The directory is set up like a typical OpenFOAM case and has the following contents (explore using the tree command on Ubuntu):

```
/opt/openfoam9/tutorials/incompressible/simpleFoam/airFoil2D/
|-- 0
|    |-- U
|    |-- nuTilda
|    |-- nut
|    `-- p
|-- Allclean
```

```
|-- Allrun
|-- constant
|    |-- momentumTransport
|    |-- polyMesh
|    |    |-- boundary
|    |    |-- cells
|    |    |-- faces
|    |    |-- neighbour
|    |    |-- owner
|    |    `-- points
|    `-- transportProperties
`-- system
     |-- controlDict
     |-- fvSchemes
     `-- fvSolution
```

Let us explore some of these files, as this will give you an understanding of any OpenFOAM case. The folder called 0 represents the initial conditions (as in, time step 0) for these key quantities we will be solving for:

- **Velocity (U)**

- **Pressure (p)**

What do these files look like? Let's take a look at the U (velocity) file:

```
FoamFile
{
    format      ascii;
    class       volVectorField;
    object      U;
}
dimensions      [0 1 -1 0 0 0 0];
internalField   uniform (25.75 3.62 0);
boundaryField
{
    inlet
    {
        type            freestreamVelocity;
```

```
        freestreamValue $internalField;
    }

    outlet
    {
        type                freestreamVelocity;
        freestreamValue $internalField;
    }

    walls
    {
        type                noSlip;
    }

    frontAndBack
    {
        type                empty;
    }
}
```

As we can see here, the file defines the dimensions of the CFD domain and the free stream velocity, along with the inlet, outlet, and wall boundary conditions.

The Airfoil2D folder also contains a folder called constant; this folder contains files specific to the CFD mesh that we will be creating. The momentumTransport file defines the kind of models to be used to solve this problem:

```
simulationType RAS;
RAS
{
    model           SpalartAllmaras;
    turbulence      on;
    printCoeffs     on;
}
```

Here, we use the SpalartAllmaras turbulence model under the **Reynolds-Averaged Flow (RAF)** type. For more information about this, please visit https://www.openfoam.com/documentation/guides/latest/doc/guide-turbulence-ras-spalart-allmaras.html.

The `boundary` file inside the `polyMesh` folder contains definitions of the walls themselves; this is to let the simulation know what a surface *inlet* or *wall* represents. There are several other files in the `polyMesh` folder that we will not explore in this section.

Inside the `System` folder, the `controlDict` file defines what applications to run for this case. OpenFOAM contains over 200 compiled applications; many of these are solvers and preprocessing and post-processing for the code.

Finally, we get to one of the most important files you will find in any OpenFOAM case: the `Allrun` executable. The `Allrun` file is a shell script that runs the steps we defined earlier for every typical CFD application in order – to import the geometry, create a mesh, solve the CFD problem, and post-process results.

Depending on the output intervals defined in your `ControlDict` file, several output folders will be output in the same directory that correspond to different time stamps in the simulation. The CFD solvers will solve the problem until they converge, or until a maximum number of time steps is reached. The output folders will look similar to the timestep `0` folder that we created earlier. To visualize these results, we use a tool called `ParaView`:

1. First, let us look at the mesh we have created (see *Figure 11.14*). The executables included within OpenFOAM that are responsible for creating this mesh are `blockmesh` and `snappyhexmesh`. You can also run these commands manually instead of running the `Allrun` file.

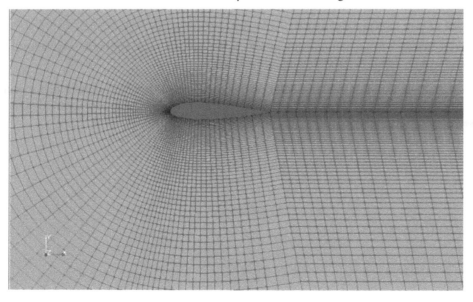

Figure 11.14 – Mesh for the Airfoil 2D case in OpenFOAM

2. Great – after solving the problem using the `SimpleFoam` executable, let us take a look at the pressure distribution around the airfoil (see *Figure 11.15*):

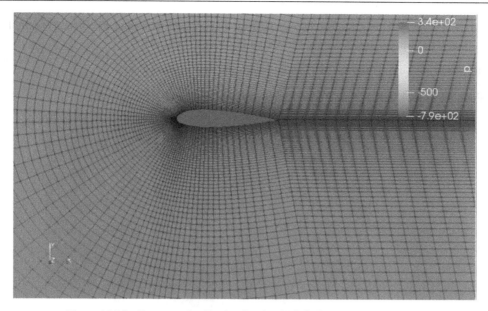

Figure 11.15 – Pressure distribution for the Airfoil 2D case in OpenFOAM

3. Lastly, we can use `ParaView` to visualize the velocity distribution, along with streamlines (see *Figure 11.16*):

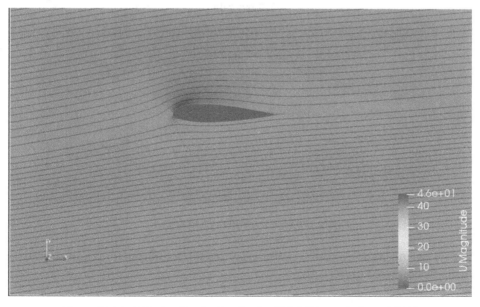

Figure 11.16 – Velocity distribution for the Airfoil 2D case in OpenFOAM

Note that these plots are post-processed by initializing `ParaView` with the `paraFoam` executable, which automatically understands output formatted by OpenFOAM cases.

Let us now look at a slightly more complicated case – the flow around a car:

1. First, let us look at the geometry of the car (*Figure 11.17* and *Figure 11.18*):

Figure 11.17 – Car geometry (perspective view)

Figure 11.18 – Car geometry (side view)

2. Next, we can use the `blockmesh` and `snappyhexmesh` commands to create the CFD mesh around this car (see *Figure 11.19*):

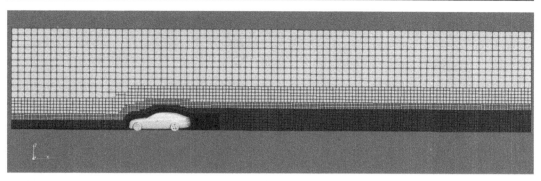

Figure 11.19 – Mesh created for the car case

3. We can then run the `Allrun` file to solve the problem. Finally, we will visualize the output (*Figures 11.20* and *Figure 11.21*):

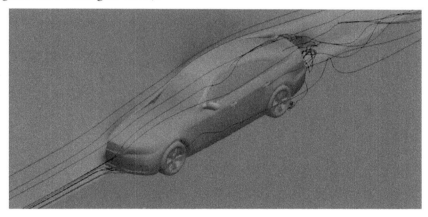

Figure 11.20 – Streamlines (black) and pressure distribution (perspective view) created for the car case

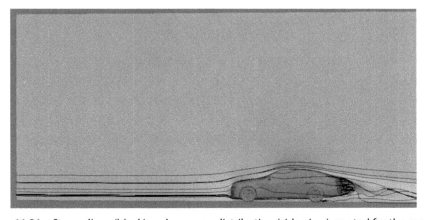

Figure 11.21 – Streamlines (black) and pressure distribution (side view) created for the car case

The files needed for the following cases can be found in the ZIP files provided in the GitHub repository here: `https://github.com/PacktPublishing/Applied-Machine-Learning-and-High-Performance-Computing-on-AWS/tree/main/Chapter11/runs`.

In the next section, we will discuss some advancements in the CFD field related to using ML and deep learning with CFD tools.

Discussing how ML can be applied to CFD

CFD, being a field that has been around for decades, has matured to be very useful to companies in various domains and has also been implemented at scale using cloud providers. Recent advances in ML have been applied to CFD, and in this section, we will provide readers with pointers to articles written about this domain.

Overall, we see deep learning techniques being applied in two primary ways:

- Using deep learning to map inputs to outputs. We explored the flow over an airfoil in this chapter and visualized these results. If we had enough input variation and saved the outputs as images, we could use **autoencoders** or **Generative Adversarial Networks (GANs)** to generate these images. As an example, the following paper uses GANs to predict flows over airfoils using sparse data: `https://www.sciencedirect.com/science/article/pii/S1000936121000728`. As we can see in *Figure 11.22*, the flow fields predicted by CFD and the GAN are visually very similar:

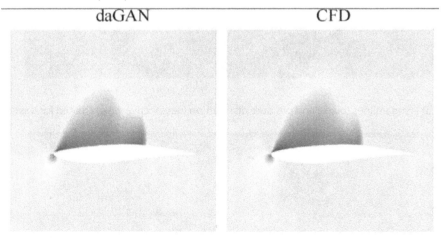

Figure 11.22 – Pressure distribution generated by a trained GAN (left) and CFD (right)

Similarly, Autodesk trained a network with over 800 examples of cars and can instantaneously predict the flow and drag of a new car body: `https://dl.acm.org/doi/10.1145/3197517.3201325` (see *figure 11.23*).

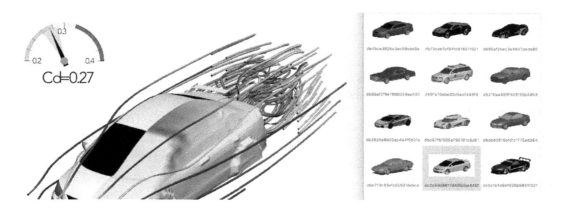

Figure 11.23 – Flow field and drag coefficient being predicted instantaneously for various car shapes

- The second general type of innovation is not just mapping inputs to outputs, but actually using ML techniques as part of the CFD solver itself. For example, NVIDIA's SIMNET (https://arxiv.org/abs/2012.07938) paper describes how deep learning can be used to model the actual **Partial Differential Equations (PDEs)** that define the fluid flow and other physical phenomena. See *Figure 11.24* for example results from SIMNET for flow over a heat sink. Parameterized training runs are faster than commercial and open source solvers, and inference for new geometries are instantaneous.

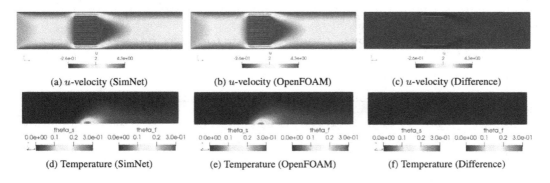

Figure 11.24 – Velocity (top row) and temperature (bottom row)
comparisons of OpenFOAM versus SIMNET from NVIDIA

Let's summarize what you have learned in this chapter next.

Summary

In this chapter, we provided a high-level introduction to the world of CFD, and then explored multiple ways to use AWS to solve CFD problems (using ParallelCluster and CFD Direct on EC2). Finally, we discussed some recent advancements connecting the field of CFD to ML. While it is out of the scope of this book to go into much more detail regarding CFD, we hope that the readers are inspired to dive deeper into the topics explored here.

In the next chapter, we will focus on genomics applications using HPC. Specifically, we will talk about drug discovery and do a detailed walk-through of a protein structure prediction problem.

References

- *AWS Batch*: https://docs.aws.amazon.com/parallelcluster/latest/ug/aws-services-v3.html#aws-batch-v3

- *AWS CloudFormation*: https://docs.aws.amazon.com/parallelcluster/latest/ug/aws-services-v3.html#aws-services-cloudformation-v3

- *Amazon CloudWatch*: https://docs.aws.amazon.com/parallelcluster/latest/ug/aws-services-v3.html#amazon-cloudwatch-v3

- *Amazon CloudWatch Logs*: https://docs.aws.amazon.com/parallelcluster/latest/ug/aws-services-v3.html#amazon-cloudwatch-logs-v3

- *AWS CodeBuild*: https://docs.aws.amazon.com/parallelcluster/latest/ug/aws-services-v3.html#aws-codebuild-v3

- *Amazon DynamoDB*: https://docs.aws.amazon.com/parallelcluster/latest/ug/aws-services-v3.html#amazon-dynamodb-v3

- *Amazon Elastic Block Store*: https://docs.aws.amazon.com/parallelcluster/latest/ug/aws-services-v3.html#amazon-elastic-block-store-ebs-v3

- *Amazon Elastic Container Registry*: https://docs.aws.amazon.com/parallelcluster/latest/ug/aws-services-v3.html#amazon-elastic-container-registry-ecr-v3

- *Amazon EFS*: https://docs.aws.amazon.com/parallelcluster/latest/ug/aws-services-v3.html#amazon-efs-v3

- *Amazon FSx for Lustre*: https://docs.aws.amazon.com/parallelcluster/latest/ug/aws-services-v3.html#amazon-fsx-for-lustre-v3

12
Genomics

Genomics is the study of an organism's genome or genetic material. In humans, genetic materials are stored in the form of **Deoxyribonucleic Acid** (**DNA**). These are the instructions that make up a human being, and 99.9% of the genomes in humans are identical and only 0.1% is different, which accounts for the differences in physical characteristics, such as eye color. Most of these variations are harmless, but some variants can cause health conditions, such as sickle cell anemia. Therefore, the analysis of such information can be used to predict or prevent a disease or provide personalized treatment, also known as **precision medicine**. There are four chemical bases present in DNA, namely **adenine** (**A**), **thymine** (**T**), **cytosine** (**C**), and **guanine** (**G**). They always bond in a particular manner; for example, adenine will always bond with thymine, and cytosine with guanine. The combination of these chemical bases is what makes up a DNA sequence.

Sequencing is at the heart of genomics. To understand what it means, the **Human Genome Project** (`https://www.ncbi.nlm.nih.gov/pmc/articles/PMC6875757/pdf/arhw-19-3-190.pdf`) was started in 1989 with the objective of sequencing one human genome within 15 years. This was completed in 12 years in 2001 and involved thousands of scientists. With the development of next-generation sequencing technology, the whole human genome can now be generated in about a day. A single human genome is around 3 billion base pairs long; a similar size has been seen for other organisms such as a mouse or a cow.

Since the time and cost of generating the genome sequence have significantly dropped, it has led to the generation of an enormous amount of data. So, in order to analyze this magnitude of data, we need powerful machines and a large amount of cost-effective storage. The good news is that DNA sequencing data is publicly available, and one of the largest repositories is the **National Center for Biotechnology Information** (**NCBI**). We can use statistical and **machine learning** (**ML**) models to gain insights from the genome data, which can be compute intensive. This poses two major challenges: big data and a massive ML model are required to make predictions, such as the prediction of promoters or predicting the masked DNA sequence.

Therefore, this chapter will help navigate these challenges by covering the following topics:

- Managing large genomics data on AWS
- Designing architecture for genomics
- Applying ML to genomics

Technical requirements

You should have the following prerequisites before getting started with this chapter:

- A web browser (for the best experience, it is recommended that you use a Chrome or Firefox browser)
- Access to the AWS account that you used in *Chapter 5, Data Analysis*
- Access to the SageMaker Studio development environment that we created in *Chapter 5, Data Analysis*
- Example Jupyter notebooks for this chapter are provided in the companion GitHub repository (`https://github.com/PacktPublishing/Applied-Machine-Learning-and-High-Performance-Computing-on-AWS/tree/main/Chapter06`)

Managing large genomics data on AWS

Apart from the large size of the genomics dataset, other challenges for managing it include discoverability, accessibility, availability, and storing it in a storage system that allows for scalable data processing while keeping the critical data safe. *The responsible and secure sharing of genomic and health data is key to accelerating research and improving human health*, is a stated objective of the **Global Alliance for Genomics and Health** (**GA4GH**). This approach requires two important things: one is a deep technical understanding of the domain, and the second is access to compute and storage resources. You can also find many genomics datasets hosted by AWS on the **Registry of Open Data on AWS** (`https://registry.opendata.aws/`).

Before you can begin any processing on the genomics dataset using cloud services, you need to make sure that it's transferred and stored on the AWS cloud. For storing data, we recommend using **Amazon Simple Storage Services** (**Amazon S3**), as genomics data produced by next-generation sequencers are persisted in files, and a lot of genomic data analysis tools also take files as inputs and write the output back as files. For example, using an ML model for data analysis might involve taking large DNA sequence files as input and storing the inference or prediction results in a file, for which Amazon S3 makes a natural fit.

You can store genomics data securely by enabling server-side encryption with either **Amazon S3-managed encryption keys** (**SSE-S3**) or **AWS Key Management Service** (**AWS KMS**) keys. Moreover, Amazon S3 also allows you to enable the data life cycle by storing the infrequently accessed data in the **Amazon S3 Standard-Infrequent Access (S3 Standard-IA)** class tier or archiving the data to a low-cost storage option, such as **Amazon S3 Glacier Deep Archive** when the data is not in use to significantly reduce the cost. This pattern is discussed in detail in the *Tiered storage for cost optimization* section of *Chapter 4, Data Storage*.

For transferring genomics data to Amazon S3, you can use the AWS DataSync service, as discussed in *Chapter 2, Data Management and Transfer*.

Let's take a closer look at the detailed architecture for applying ML models to the genomics dataset.

Designing architecture for genomics

In this section, we will describe a sample reference architecture for transferring, storing, processing, and gaining insights on genomics datasets on the AWS cloud, in a secure and cost-effective manner. *Figure 12.1* shows a sample genomics data processing workflow:

Figure 12.1 – Genomics data processing workflow

Figure 12.1 shows the following workflow:

1. A scientist or a lab technician will collect sample genomic data, for example, skin cells, prepare it in a lab, and then load it into a sequencer.

2. The sequencer will then generate a sequence, which might be short DNA fragments. These are usually called **reads** because you are reading DNA.

3. The DNA sequence is stored in an on-premises data storage system.

4. The AWS DataSync service will then transfer the genomic data securely to the cloud; for further details, refer to *Chapter 2, Data Management and Transfer*.

5. The raw genomic data is then stored on Amazon S3. You can use AWS Analytics tools for data processing.

6. Amazon Genomics CLI is a purpose-built open source tool for processing raw genomics data in the cloud at a petabyte scale. For details, please refer to this link: `https://aws.amazon.com/genomics-cli/`.

7. Optionally, we recommend storing the processed genomics data on **Amazon Feature Store**, which is a fully managed service for storing, sharing, versioning, and managing ML features for training and inference to enable the reuse of features across ML applications.

8. You can add granular access control policies on genomics data stored on Amazon S3 or Amazon Feature Store by using the **AWS Lake Formation** service, based on your business requirements. For details on AWS Lake Formation, refer to this link: `https://aws.amazon.com/lake-formation`.

9. Once the data is processed and stored on either Amazon Feature Store or Amazon S3, you can run ML models such as **DNABERT** (`https://www.biorxiv.org/content/10.1101/2020.09.17.301879v1`) using **Amazon SageMaker** to gain further insights or to predict the masked DNA sequence. The ML model can take a batch of genomic data, make inferences, and store the results back in Amazon S3.

10. Additionally, you can archive the unused data to Amazon S3 Glacier Deep Archive to have significant cost savings on data storage.

> **Note**
>
> A detailed discussion on Amazon Genomics CLI, AWS Lake Formation, and Amazon Feature Store is out of scope for this chapter; however, we will use the DNABERT model in the *Applying ML to genomics* section of this chapter.

Let's learn how to apply the ML model to genomics applications and predict masked sequences in a DNA sequence using a pretrained ML model.

Applying ML to genomics

Before we dive into ML model details, let's first understand the genomic data, which is stored as DNA in every organism. There are four chemical bases present in DNA, namely **Adenine (A)**, **Thymine (T)**, **Cytosine (C)** and **Guanine (G)**. They always bond in particular manner for example, Adenine will always bond with Thymine, and Cytosine with Guanine. The combination of these chemical bases is what makes up a DNA sequence, represented by the letters A, T, C, and G. A 20-length example of a DNA sequence is `ACTCCACAGTACCTCCGAGA`. A single complete sequence of the human genome is around 3 billion **base pairs** (**bp**) long and takes about 200 GB of data storage (`https://www.science.org/doi/10.1126/science.abj6987`).

However, for analyzing the DNA sequence, we don't need the complete human genome sequence. Usually, we analyze a part of the human DNA; for example, to determine hair growth or skin growth, a lab technician will take a small section of human skin and prepare it to run through the next-generation sequencer, which will then read the DNA and generate the DNA sequence, which are short fragments of DNA. ML models can be used for various tasks, such as DNA classification, **promoter recognition**, interpretation of structural variation in human genomes, precision medicine, cancer research, and so on.

In this section, we will showcase how to fine-tune the DNABERT model using Amazon SageMaker for the proximal promoter recognition task. DNABERT is based on the BERT model fine-tuned on DNA sequences, as outlined in the research paper *Supervised promoter recognition: a benchmark framework* (`https://bmcbioinformatics.biomedcentral.com/track/pdf/10.1186/s12859-022-04647-5.pdf`). Therefore, let's take an example of deploying a pretrained DNABERT model for promoter recognition using DNA sequence data on Amazon SageMaker service.

Amazon SageMaker is a fully managed service by AWS to assist ML practitioners in building, training, and deploying ML models. Although it provides features and an integrated development environment for each phase of ML, it is highly modular in nature, meaning if you already have a trained model, you can use the **SageMaker hosting** features to deploy the model for performing inference/predictions on your model. For details on the various types of deployment options provided by SageMaker, refer to *Chapter 7, Deploying Machine Learning Models at Scale*.

We will deploy a version of the DNABERT pretrained model provided by **Hugging Face**, an AI community with a library of 65K+ pretrained transformer models. Amazon SageMaker provides first-party deep learning containers for Hugging Face for both training and inference. These containers include Hugging Face pretrained transformer models, tokenizers, and dataset libraries. For a list of all the available containers, you can refer to this link: `https://github.com/aws/deep-learning-containers/blob/master/available_images.md`. These containers are regularly maintained and updated with security patches and remove undifferentiated heavy lifting for ML practitioners.

With a few lines of configuration code, you can deploy a pretrained model in the Hugging Face library on Amazon SageMaker, and start making predictions using your model. Amazon SageMaker provides a lot of features for deploying ML models. For example, in the case of **real-time inference**, when you choose to deploy the model as an API, set up an **autoscaling policy** to scale up and down the number of instances on which the model is deployed based on the number of invocations on the model. Additionally, you can do **blue/green deployment**, add **security guardrails**, **auto-rollback**, and so on, using Amazon SageMaker hosting features. For details on deploying models for inference, refer to this link: `https://docs.aws.amazon.com/sagemaker/latest/dg/deploy-model.html`. Now that we have understood the benefits of using Amazon SageMaker for deploying models and Hugging Face integration, let's see how we can deploy a pretrained DNABERT model for promoter recognition.

> **Note**
> The full code for deploying the model is available on GitHub: `https://github.com/PacktPublishing/Applied-Machine-Learning-and-High-Performance-Computing-on-AWS/blob/main/Chapter12/dnabert.ipynb`.

We need to follow three steps for deploying pretrained transformer models for real-time inference provided by the Hugging Face library on Amazon SageMaker:

1. Provide model hub configuration where we supply the Hugging Face model ID and the task – in our case, text classification.

2. Create a `HuggingFaceModel` class provided by the SageMaker API, where we provide parameters such as the transformer version, PyTorch version, Python version, hub configuration, and role.

3. Finally, we use the `deploy()` API, where we supply the number of instances and the type of instance on which to deploy the model.

The following code snippet showcases the three steps that we just outlined:

```
...
from sagemaker.huggingface import HuggingFaceModel
import sagemaker

role = sagemaker.get_execution_role()
# Step 1: Hub Model configuration. https://huggingface.co/
models
hub = {
    'HF_MODEL_ID':'AidenH20/DNABERT-500down',
    'HF_TASK':'text-classification'
}

# Step 2: create Hugging Face Model Class
huggingface_model = HuggingFaceModel(
    transformers_version='4.17.0',
    pytorch_version='1.10.2',
    py_version='py38',
    env=hub,
    role=role,
)

# Step 3: deploy model to SageMaker Inference
predictor = huggingface_model.deploy(
    initial_instance_count=1, # number of instances
    instance_type='ml.m5.xlarge' # ec2 instance type
```

```
)

...
```

Using this code snippet, we basically tell SageMaker to deploy the Hugging Face model provided in `'HF_MODEL_ID'` for the task mentioned in `'HF_TASK'`; in our case, `text classification`, as we want to classify promoter regions by providing a DNA sequence. The `HuggingFaceModel` class defines the container on which the model will be deployed. Finally, the `deploy()` API launches the Hugging Face container defined by the `HuggingFaceModel` class and loads the model provided in the hub configuration to the initial number of instances and type of instances provided by the ML practitioner.

> **Note**
>
> The number of instances on which the model is deployed as an API can be updated even after the model is deployed.

Once the model is deployed, you can then use the `predict()` API provided by SageMaker to make inferences or predictions on the model, as shown in the following code snippet:

```
...
dna_sequence = 'CTAATC TAATCT AATCTA ATCTAG TCTAGT CTAGTA
TAGTAA AGTAAT GTAATG TAATGC AATGCC ATGCCG TGCCGC GCCGCG CCGCGT
CGCGTT GCGTTG CGTTGG GTTGGT TTGGTG TGGTGG GGTGGA GTGGAA TGGAAA
GGAAAG GAAAGA AAAGAC AAGACA AGACAT GACATG ACATGA CATGAC ATGACA
TGACAT GACATA ACATAC CATACC ATACCT TACCTC ACCTCA CCTCAA CTCAAA
TCAAAC CAAACA AAACAG AACAGC ACAGCA CAGCAG AGCAGG GCAGGG CAGGGG
AGGGGG GGGGGC GGGGCG GGGCGC GGCGCC GCGCCA CGCCAT GCCATG CCATGC
CATGCG ATGCGC TGCGCC GCGCCA CGCCAA GCCAAG CCAAGC CAAGCC AAGCCC
AGCCCG GCCCGC CCCGCA CCGCAG CGCAGA GCAGAG CAGAGG AGAGGG GAGGGT
AGGGTT GGGTTG GGTTGT GTTGTC TTGTCC TGTCCA GTCCAA TCCAAC CCAACT
CAACTC AACTCC ACTCCT CTCCTA TCCTAT CCTATT CTATTC TATTCC ATTCCT'
predictor.predict({
    'inputs': dna_sequence
})
...
```

The output will be the label with the highest probability. In our case, it's either LABEL_0 or LABEL_1, denoting the absence or presence of a promoter region in a DNA sequence.

> **Note**
>
> The preceding code deploys the model as an API on a long-running instance, so if you are not using the endpoint, please make sure to delete it; otherwise, you will be charged for it.

You can also see the endpoint details on SageMaker Studio by clicking on an orange triangle icon (**SageMaker resources**) in the left navigation panel and selecting **Endpoints**, as shown in the following screenshot:

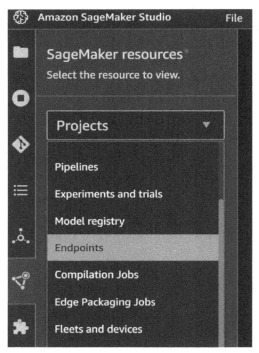

Figure 12.2 – Accessing Endpoints on SageMaker Studio

This will show all the SageMaker endpoints (models deployed for real-time inference as an API). Double-clicking on the endpoint name will show you the details by calling the `DescribeEndpoint()` API behind the scenes. The SageMaker Studio UI shows you a lot of options, such as **Test inference**, **Data quality**, **Model quality**, **Model explainability**, **Model bias**, **Monitoring job history**, and **AWS settings**. The data is populated on these tabs based on the features you have enabled; for example, to understand the data and model quality, you need to enable the model monitoring feature of SageMaker, which monitors the models deployed as real-time endpoints on a schedule set by you, compares it with baseline statistics, and stores the report in S3. For details on model monitoring, refer to this link: `https://docs.aws.amazon.com/sagemaker/latest/dg/model-monitor.html`.

The **AWS settings** tab, on the other hand, will always be populated with model endpoint metadata, such as the endpoint name, type, status, creation time, last updated time, **Amazon Resource Name** (**ARN**), endpoint runtime settings, endpoint configuration settings, production variants (in case you have more than one variant of the same model), instance details (type and number of instances), model name, and lineage, as applicable. *Figure 12.3* shows some of the metadata associated with the endpoint:

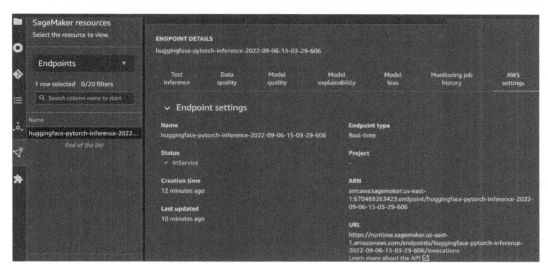

Figure 12.3 – SageMaker endpoint details

Also, if you quickly want to test your model from the SageMaker Studio UI, you can click on **Test inference**, provide a payload (input request) in JSON format in **JSON editor**, as shown in *Figure 12.4*, and quickly see the response provided by the model:

Figure 12.4 – Test inference from the SageMaker Studio UI

Now that we understand how pretrained models on the Hugging Face model library can be deployed and tested on Amazon SageMaker, let's take another example of how to fine-tune the pretrained model in the Hugging Face library, and deploy the fine-tuned model. For this section, we will use a BERT-based model trained on protein sequences known as **ProtBERT**, published in this research paper: `https://www.biorxiv.org/content/10.1101/2020.07.12.199554v3`.

The study of protein structure, functions, and interactions is called **proteomics**, which takes the help of genomic studies as proteins are the functional product of the genome. Both proteomics and genomics are used to prevent diseases and are an active part of drug discovery. Although there are a lot of tasks that contribute to drug discovery, protein classification, and secondary structure identification play a vital role. In the following section, we will understand how to fine-tune a large protein model (the Hugging Face library) to predict the secondary structure of a protein using the distributed training features of Amazon SageMaker.

Protein secondary structure prediction for protein sequences

Protein sequences are made up of 20 essential **amino acids**, each represented by a capital letter. They combine to form a protein sequence, which you can use to do protein classification or predict the secondary structure of the protein, among other tasks. Protein sequences assume a particular 3D structure based on constraints, which are optimized for undertaking a particular function. You can think of these constraints as rules of grammar or meaning in natural language, which allows us to use **natural language processing** (NLP) techniques to protein sequences.

In this section, we will focus on fine-tuning the `prot_t5_xl_uniref50` model, which has around *11 billion parameters*, and you can use the same training script to fine-tune a smaller `prot_bert_bfd` model, which has close to *420 million parameters*, with different configuration to accommodate the size of the model. The code for fine-tuning the `prot_t5_xl_uniref50` model is provided in the GitHub repository: `https://github.com/PacktPublishing/Applied-Machine-Learning-and-High-Performance-Computing-on-AWS/tree/main/Chapter12`.

To create a SageMaker training job using the model in the Hugging Face library, we will need a Hugging Face estimator from SageMaker SDK. The estimator class will handle all the training tasks based on the configuration that we provide. For example, to train a Hugging Face model with the basic configuration, we will need to provide the following parameters:

- `entry_point`: This is where we will specify the training script for training the model.
- `source_dir`: This is the name of the folder where the training script or other helper files will reside.
- `instance_type`: This is the type of machine in the cloud where the training script will run.
- `instance_count`: This is the number of machines in the cloud that will be launched for running the training job. If the count is greater than 1, then it will automatically launch a cluster.

- `transfomer_version`, `pytorch_version`, `py_version`: These determine the version of transformers, PyTorch, and Python that will be present in the container. Based on the value of these parameters, SageMaker will fetch the appropriate container, which will be provisioned onto the instances (machines) in the cloud.

- `hyperparameters`: This defines the named arguments that will be passed to the training script.

The following code snippet formalizes these parameters into a SageMaker training job:

```
...
huggingface_estimator = HuggingFace(entry_point='train.py',
                        source_dir='./code',
                        instance_type='ml.g5.12xlarge',
                        instance_count=1,
                        role=role,
                        transformers_version='4.12',
                        pytorch_version='1.9',
                        py_version='py38',
                        hyperparameters=hyperparameters)
...
```

Once the estimator is defined, you can then provide the S3 location (path) of your data to start the training job. SageMaker provides some very useful environment variables for a training job, which include the following:

- `SM_MODEL_DIR`: This provides the path where the training job will store the model artifacts, and once the job finishes, the stored model in this folder is directly uploaded to the S3 output location.

- `SM_NUM_GPUS`: This represents the number of GPUs available to the host.

- `SM_CHANNEL_XXXX`: This represents the input data path for the specified channel. For example, `data`, in our case, will correspond to the `SM_CHANNEL_DATA` environment variable, which is used by the training script, as shown in the following code snippet:

```
...
# starting the train job with our uploaded datasets as
input
huggingface_estimator.fit({'data': data_input_path})
...
```

When we call the `fit()` method on the Hugging Face estimator, SageMaker will first provision the ephemeral compute environment, and copy the training script and data to the compute environment. It will then kick off the training, save the trained model to the S3 output location provided in the estimator class, and finally, tear down all the resources so that users only pay for the amount of time the training job was running for.

> **Note**
>
> For more information on the SageMaker Hugging Face estimator, as well as how to leverage the SageMaker SDK to instantiate the estimator, please see the AWS documentation (`https://docs.aws.amazon.com/sagemaker/latest/dg/hugging-face.html`), and the SageMaker SDK documentation (`https://sagemaker.readthedocs.io/en/stable/frameworks/huggingface/sagemaker.huggingface.html#hugging-face-estimator`).

We will extend this basic configuration to fine-tune `prot_t5_xl_uniref50` with *11 billion parameters* using the distributed training features of SageMaker.

Before we do a deeper dive into the code, let's first understand some of the concepts that we will leverage for training the model. Since it's a big model, the first step is to get an idea of the size of the model, which will help us determine whether it will fit in a single GPU memory or not. The SageMaker model parallel documentation (`https://docs.aws.amazon.com/sagemaker/latest/dg/model-parallel-intro.html`) gives a good idea of estimating the size of the model. For a training job that uses **automatic mixed precision (AMP)** with a **floating-point 16-bit (FP16)** size, using Adam optimizers, we can calculate the memory required per parameter using the following breakdown, which comes to about 20 bytes:

- An FP16 parameter ~ 16 bits or 2 bytes
- An FP16 gradient ~ 16 bits or 2 bytes
- An FP32 optimizer state ~ 64 bits or 8 bytes based on the Adam optimizers
- An FP32 copy of parameter ~ 32 bits or 4 bytes (needed for the **optimizer apply (OA)** operation)
- An FP32 copy of gradient ~ 32 bits or 4 bytes (needed for the OA operation)

Therefore, our model with 11 billion parameters will require more than 220 GB of memory, which is larger than the typical GPU memory currently available on a single GPU. Even if we are able to get such a machine with GPU memory greater than 220 GB, it will not be cost-effective, and we won't be able to scale our training job. Another constraint to understand here is the batch size, as we will need at least one batch of data in the memory to begin training. Using a smaller batch size will drive down the GPU utilization and degrade the training efficiency, as the model might not be able to converge.

Hence, we will have to partition our model into multiple GPUs, and in order to increase the batch size, we will also need to shard the data. So, we will use a hybrid approach that will utilize both data parallel and model parallel approaches. This approach has been explained in detail in *Chapter 6, Distributed Training of Machine Learning Models*.

Since our model size is 220 GB, we will have to use a mechanism in the training job, which will optimize it for memory in order to avoid **out of memory (OOM)** errors. For memory optimization, the SageMaker model parallel library provides two types of model parallelism, namely, pipeline parallelism and tensor parallelism, along with memory-saving techniques, such as activation checkpointing, activation offloading, and optimizer state sharding. Let's understand each of these terms:

- **Pipeline parallelism**: This means partitioning the model into different GPUs. This goes as a parameter to the training job, and based on the parallelism degree provided, the library will create different partitions. For example, we are training the model with 4 GPUs, and if the `partitions` parameter value provided in the configuration is 2, then the model will be divided into 2 partitions across 4 GPUs, meaning it will have 2 model replicas. Since, in this case, the number of GPUs is greater than the number of partitions, we will have to enable **distributed data parallel** by setting the `ddp` parameter to `true`. Otherwise, the training job will give an error. Pipeline parallelism with `ddp` enabled is illustrated in *Figure 12.5*:

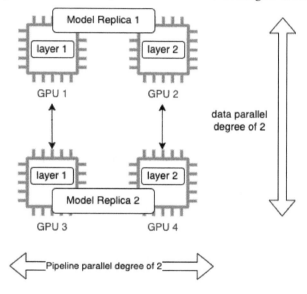

Figure 12.5 – Showcasing the hybrid approach with model and data parallelism

- **Tensor Parallelism**: This applies the same concept as pipeline parallelism and goes a step further to divide the largest layer of our model and places it on different nodes. This concept is illustrated in *Figure 12.6*:

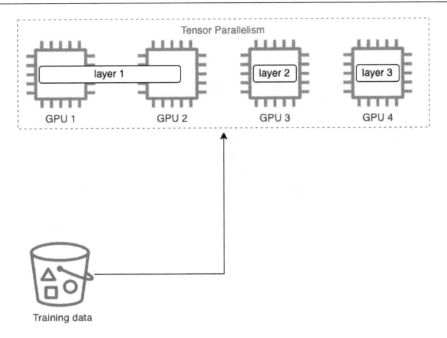

Figure 12.6 – Tensor parallelism shards tensors (layers) of the model across GPUs

- **Activation checkpointing**: This helps in reducing memory usage by clearing out the activations of layers and computing it again during the backward pass. For any deep learning model, the data is first passed through the intermediary layers in the forward pass, which compute the outputs, also known as activations. These activations need to be stored as they are used for computing gradients during the backward pass. Now, storing activations for a large model in memory can increase memory usage significantly and can cause bottlenecks. In order to overcome this issue, an activation checkpointing or gradient checkpointing technique comes in handy, which clears out the activations of intermediary layers.

> **Note**
> Activation checkpointing results in extra computation time in order to reduce memory usage.

- **Activation offloading**: This uses activation checkpointing, where it only keeps a few activations in the GPU memory during the model training. It offloads the checkpointed activations to CPU memory during the forward pass, which are loaded back to the GPU during the backward pass of a particular micro-batch of data.

- **Optimizer state sharding**: This is another useful memory-saving technique that partitions the state of the optimizers described by the set of weights across the data parallel device groups. It can be used only when we are using a stateful optimizer, such as Adam or FP16.

> **Note**
>
> Since optimizer state sharding partitions the optimizer state across the data parallel device groups, it will only become useful if the data parallel degree is greater than one.

Another important concept to understand with model parallelism is the **micro-batch**, which is a smaller subset of the batch of data. During training, you pass a certain number of records of data forward and backward through the layers, known as a batch or sometimes a mini-batch. A full pass through your dataset is called an **epoch**. SageMaker model parallelism shards the batch into smaller subsets, which are called micro-batches. These micro-batches are then passed through the **pipeline scheduler** to increase GPU utilization. The pipeline scheduler is explained in detail in *Chapter 6, Distributed Training of Machine Learning Models*.

So, now that we understand the memory-saving techniques used by the SageMaker model parallel library let's see how we can use them in the code. The following code snippet wraps all the memory-saving techniques together in a simple manner:

```
...
# configuration for running training on smdistributed Model
Parallel
mpi_options = {
    "enabled": True,
    "processes_per_host": 8,
}
smp_options = {
    "enabled":True,
    "parameters": {
        "microbatches": 1,
        "placement_strategy": "cluster",
        "pipeline": "interleaved",
        "optimize": "memory",
        "partitions": 4,
        "ddp": True,
        # "tensor_parallel_degree": 2,
        "shard_optimizer_state": True,
        "activation_checkpointing": True,
        "activation_strategy": "each",
        "activation_offloading": True,
    }
}
```

```
distribution = {
    "smdistributed": {"modelparallel": smp_options},
    "mpi": mpi_options
}
...
```

All the memory-saving techniques are highlighted in the code snippet, where you first have to make sure to set the optimize parameter to memory. This instructs the SageMaker model splitting algorithm to optimize for memory consumption. Once this is set, then you can simply enable other memory-saving features by setting their value to true.

You will then supply the distribution configuration to the HuggingFace estimator class, as shown in the following code snippet:

```
...
huggingface_estimator = HuggingFace(entry_point='train.py',
                    source_dir='./code',
                    instance_type='ml.g5.48xlarge',
                    instance_count=1,
                    role=role,
                    transformers_version='4.12',
                    pytorch_version='1.9',
                    py_version='py38',
                    distribution= distribution,
                    hyperparameters=hyperparameters)
huggingface_estimator.fit({'data': data_input_path})
...
```

As you can see, we also provide the train.py training script as entry_point to the estimator. In *Chapter 6*, *Distributed Training of Machine Learning Models*, we understood that when we are using model parallel, we have to update our training script with SageMaker model parallel constructs. In this example, since we are using the HuggingFace estimator and the Trainer API for model training, it has built-in support for SageMaker model parallel. So, we simply import the Hugging Face Trainer API and provide configuration related to model training, and based on the model parallel configuration provided in the HuggingFace estimator, it will invoke the SageMaker model parallel constructs during model training.

In our `train.py` script, first, we need to import the `Trainer` module, as shown in the following code snippet:

```
...
from transformers.sagemaker import SageMakerTrainingArguments
as TrainingArguments
from transformers.sagemaker import SageMakerTrainer as Trainer
...
```

Since we are training a T5-based BERT model (`prot_t5_xl_uniref50`) we will also need to import the `T5Tokenizer` and `T5ForConditionalGeneration` modules from the `transformers` library:

```
...
from transformers import AutoTokenizer,
T5ForConditionalGeneration, AutoModelForTokenClassification,
BertTokenizerFast, EvalPrediction, T5Tokenizer
...
```

The next step is to transform the protein sequences and load them into a PyTorch DataLoader. After we have the data in `DataLoader`, we will provide `TrainingArguments`, as shown in the following code snippet, which will be used by the `Trainer` API:

```
...
training_args = TrainingArguments(
    output_dir='./results',           # output directory
    num_train_epochs=2,               # total number of training
epochs
    per_device_train_batch_size=1,    # batch size per device
during training
    per_device_eval_batch_size=1,     # batch size for evaluation
    warmup_steps=200,                 # number of warmup steps
for learning rate scheduler
    learning_rate=3e-05,              # learning rate
    weight_decay=0.0,                 # strength of weight decay
    logging_dir='./logs',             # directory for storing
logs
    logging_steps=200,                # How often to print logs
    do_train=True,                    # Perform training
    do_eval=True,                     # Perform evaluation
```

```
    evaluation_strategy="epoch",        # evalute after each epoch
    gradient_accumulation_steps=32,     # total number of steps
before back propagation
    fp16=True,                          # Use mixed precision
    fp16_opt_level="02",                # mixed precision mode
    run_name="ProBert-T5-XL",          # experiment name
    seed=3,                             # Seed for experiment
reproducibility
    load_best_model_at_end=True,
    metric_for_best_model="eval_accuracy",
    greater_is_better=True,
    save_strategy="epoch",
    max_grad_norm=0,
    dataloader_drop_last=True,
    )
...
```

As you can see, `TrainingArguments` contains a list of hyperparameters, such as the number of epochs, learning rate, weight decay, evaluation strategy, and so on, which will be used for model training. For details on the different hyperparameters for the `TrainingArguments` API, you can refer to this URL: `https://huggingface.co/docs/transformers/v4.24.0/en/main_classes/trainer#transformers.TrainingArguments`. Make sure that when you are providing `TrainingArguments`, the value of `dataloader_drop_last` is set to `true`. This will make sure that the batch size is divisible by the micro-batches, and setting `fp16` to `true` will use the automatic mixed precision for model training, which also helps reduce the memory footprint, as floating-point 16 takes 2 bytes to store a parameter.

Now, we will define the `Trainer` API, which takes `TrainingArguments` and the training and validation datasets as input:

```
...
trainer = Trainer(
    model_init=model_init,              # the instantiated
Transformers model to be trained
    args=training_args,                 # training arguments,
defined above
    train_dataset=train_dataset,        # training dataset
    eval_dataset=val_dataset,           # evaluation dataset
    compute_metrics = compute_metrics,  # evaluation metrics
    )
...
```

Once the `Trainer` API has been defined, the model training is kicked off with the `train()` method, and once the training is complete, we will use the `save_model()` method to save the trained model to the specified model directory:

```
...
trainer.train()
trainer.save_model(args.model_dir)
...
```

The `save_model()` API takes the model path as a parameter, coming from the `SM_MODEL_DIR` SageMaker container environment variable and parsed as a `model_dir` variable. The model artifacts stored in this path will then be copied to the S3 path specified in the `HuggingFace` estimator, and all the resources, such as the training instance, will then be torn down so that users only pay for the duration of the training job.

> **Note**
>
> We are training a very large model, `prot_t5_xl_uniref50`, with 11 billion parameters on a very big instance, `ml.g5.48xlarge`, which is an NVIDIA A10G tensor core machine with 8 GPUs and 768 GB of GPU memory. Although we are using model parallel, the training of the model will take more than 10 hours, and you will incur a cost for it. Alternatively, you can use a smaller model, such as `prot_bert_bfd`, which has approximately 420 million parameters and is pretrained on protein sequences. Since it's a relatively smaller model that can fit into a single GPU memory, you can only use the SageMaker distributed data parallel library, as described in *Chapter 6, Distributed Training of Machine Learning Models*.

Once the model is trained, you can deploy the model using the SageMaker `HuggingFaceModel` class, as explained in the *Applying ML to genomics* main section of this chapter, or simply use the `huggingface_estimator.deploy()` API, as shown in the following code snippet:

```
...
predictor = huggingface_estimator.deploy(1, "ml.g4dn.xlarge")
...
```

Once the model is deployed, you can use the `predictor` variable to make predictions:

```
...
predictor.predict(input_sequence)
...
```

> **Note**
> Both the discussed deployment options will deploy the model for real-time inference as an API on a long-running instance. So, if you are not using the endpoint, make sure to delete it; otherwise, you will incur a cost for it.

Now that we understand the applications of ML to genomics, let's recap what we have learned so far in the next section.

Summary

In this chapter, we started with understanding the concepts of genomics and how you can store and manage large genomics data on AWS. We also discussed the end-to-end architecture design for transferring, storing, analyzing, and applying ML to genomics data using AWS services. We then focused on how you can deploy large state-of-the-art models for genomics, such as DNABERT, for promoter recognition tasks using Amazon SageMaker with a few lines of code and how you can test your endpoint using code and the SageMaker Studio UI.

We then moved on to understanding proteomics, which is the study of protein sequences, structure, and their functions. We walked through an example of predicting protein secondary structure for protein sequences using a Hugging Face pretrained model with 11 billion parameters. Since it is a large model with memory requirements greater than 220 GB, we explored various memory-saving techniques, such as activation checkpointing, activation offloading, optimizer state sharding, and tensor parallelism, provided by the SageMaker model parallel library. We then used these techniques to train our model for predicting protein structure. We also understood how SageMaker provides integration with Hugging Face and makes it simple to use state-of-the-art models, which otherwise need a lot of heavy lifting in order to train.

In the next chapter, we will review another domain, autonomous vehicles, and understand how high-performance computing capabilities provided by AWS can be used for training and deploying ML models at scale.

13
Autonomous Vehicles

Today, almost every car company is advancing technology in their cars using **Autonomous Vehicle (AV)** systems and **Advanced Driver Assistance Systems (ADAS)**. This covers everything from cruise control to several safety features and fully autonomous driving that you are all probably familiar with. If you are not familiar with these concepts, we encourage you to take the following crash course – test drive a car with fully autonomous capabilities to appreciate the technology and sophistication involved in building these kinds of systems. Companies that are currently heavily investing in AV and ADAS systems require heavy computational resources to test, simulate, and develop related technologies before deploying them in their cars. Many companies are turning to the cloud when there is a need for on-demand, elastic compute for these large-scale applications. The previous chapters have covered storage, network, and computing, and introduced ML in general.

In this chapter, we will broadly cover what AV systems and ADAS are, and how AWS compute and ML services help with the design and deployment of AV/ADAS architectures. Specifically, we will cover the following topics:

- Introducing AV systems
- AWS services supporting AV systems
- Designing an architecture for AV systems
- ML applied to AV systems

By the end of this chapter, you will have learned about the following:

- The technology used in AVs at a high level
- AWS services that can be used to create and test software related to AVs
- How machine learning is used in the development of AVs

Technical requirements

You should have the following prerequisites before getting started with this chapter:

- Familiarity with AWS and its basic usage.

- A web browser. (For the best experience, it is recommended that you use a Chrome or Firefox browser.)

- An AWS account. (If you are unfamiliar with how to get started with an AWS account, you can go to this link: `https://aws.amazon.com/getting-started/`.)

Introducing AV systems

As we mentioned earlier, several automotive companies are already implementing ADAS and AV systems in their vehicles. As such, there is a large amount of research and development happening in the space, but this section will introduce you to key terms and concepts so we can proceed further and explore how machine learning is involved here.

First, let us discuss ADAS and AV at a high level. There are several threads of questions from people being introduced to this field leading to confusion, such as the following:

- Are AV systems and ADAS one and the same thing?

- Are ADAS contained within AV systems?

- Does a company usually develop ADAS first and then AV systems?

- Are there different levels of automation within AV systems?

Before we can answer these questions, let's drill down even further. In order to automate any part of the driving experience, whether it is for a car or a container truck, innovation in the following components becomes necessary:

- Driver assistance or self-driving hardware

- Driver assistance or self-driving software

- Data and compute services

The first step in adding these technologies is adding the right hardware. Usually, this comprises a combination of RADAR, LiDAR, and **Camera sensors**.

- **Radio Detection and Ranging**, or **RADAR**, uses radio waves to estimate the distance and velocity of objects around it. RADARs are classified based on their range as short-range, mid-range, and long-range. Shorter-range RADARs are used for functions such as parking distance assistance and blind spot monitoring. Longer-range RADARs are used for lane following, automatic braking, and so on.

- **Light Detection and Ranging**, or **LiDAR**, is similar to RADAR but uses the reflection of light from surfaces to determine the distance to objects. High-resolution LiDAR can also be used to determine the shape of objects along with **Deep Learning** (**DL**) algorithms, as we will learn later in this chapter.

- Lastly, cameras are placed around the vehicles for low-level tasks such as parking, as well as high-level ones such as fully autonomous driving (when the right DL algorithms are used along with a full camera array). Elon Musk, the CEO of Tesla, a self-driving car company with a large number of cars currently sold with self-driving capabilities, famously preferred Tesla cars to be designed with camera systems only – no RADAR or LiDAR – since he said, humans only depend on vision to drive. The following figure shows an older systems architecture for Tesla, which now heavily depends on computer vision-based systems rather than LiDAR.

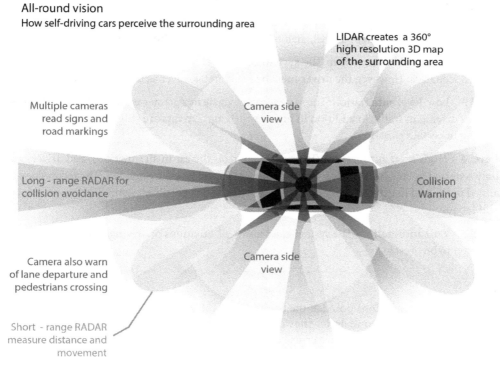

Figure 13.1 – Camera, RADAR, and LiDAR sensors mounted on a self-driving vehicle.

On a typical car chassis, you can imagine these sensors mounted, as seen in *Figure 13.1*. As we can see, there are several cameras, RADAR, and LiDAR sensors that are used to achieve various levels of autonomous driving for vehicles. *Now, what are these levels of autonomous driving?*

Along with some of the hardware (sensors) defined previously, a complex software stack is required to build and maintain the features required for AV development. These features are categorized by the **Society of Automotive Engineers** (**SAE**) and are widely adopted. The SAE defined (in 2014 and later revised) five levels of autonomous driving (see here: `https://www.sae.org/blog/sae-j3016-update`):

- **Level 0**: Features that provide warnings and limited assistance, for example, automatic emergency braking, blind spot warnings, and lane departure warnings.

- **Level 1 – Basic driver assistance**: The driver can take their feet off the pedals but needs to keep their hands on the wheel to take over, for example, lane centering or cruise control. Note that cruise control can be *adaptive* as well, maintaining a safe distance from the vehicle in front of you.

- **Level 2 – Limited automation**: The system controls steering, braking, and driving, but in limited scenarios. However, the driver must be ready to take over as needed to maintain safety.

- **Level 3 – Low-level automation**: The system can navigate in a greater number of circumstances, such as driving in traffic in addition to highway driving. Drivers are still required to take over driving in certain situations.

- **Level 4 – High-level automation**: The system controls the car in all situations, and drivers are only required to take over rarely when unknown situations are encountered. The technology is aimed to be used in driverless taxis and trucks, with the driver still present, to reduce workload and fatigue.

- **Level 5 – Full automation**: The system can handle all situations of driving, and a driver is not required to be present.

> **Note**
>
> To clarify one of the questions posed at the beginning of this chapter, ADAS may be part of a larger AV system or used standalone. They are primarily focused on lower-level automation tasks and driver aids such as adaptive cruise control or driver alertness warning systems.

The following figure shows the SAE definitions for levels of automation:

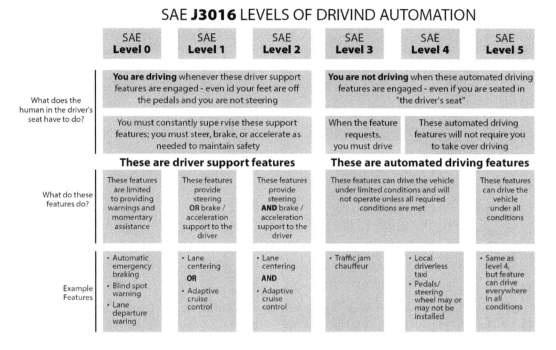

Figure 13.2 – SAE levels of automation. Source: `https://www.sae.org/blog/sae-j3016-update`

In this section, we have introduced basic concepts around AV systems at a high level, including hardware and software components required in the development of these systems. In the next section, we will take a look at AWS services that support the development of AV systems.

AWS services supporting AV systems

The development and testing of AV systems and ADAS require a cloud platform with highly scalable compute, storage, and networking. Being HPC applications, these components were covered in detail in previous chapters. As a recap, we have covered the following topics that are still relevant in this chapter:

- *Chapter 3*, *Compute and Networking* (see topics on architectural patterns and compute instances)
- *Chapter 4*, *Data Storage* (see topics on Amazon S3 and FSx for Lustre)
- *Chapter 5*, *Data Analysis and Preprocessing* (see topics on large-scale data processing)
- *Chapters 6* to *9* (covering distributed training and deployment on the cloud and at the edge)

In this section, we will highlight some more services, including the ones that we discussed in the context of AV and ADAS development. A single autonomous vehicle can generate several TB of data per day. This data is used across the AV development workflow that is discussed at the following link (`https://aws.amazon.com/blogs/architecture/field-notes-building-an-autonomous-driving-and-adas-data-lake-on-aws/`) and includes the following:

- Data acquisition and ingestion
- Data processing and analytics
- Labeling
- Map development
- Model and algorithm development
- Simulation
- Verification and validation
- Deployment and orchestration

With ever-expanding data sizes, customers look for scalable, virtually unlimited-capacity data stores at the center of all the preceding activities. For example, Lyft Level 5 manages **Petabytes** (**PB**) of storage data from car sensors in **Amazon S3**. Amazon S3 is also central to the concept of building a **data lake** for applications in the AV space, which will be discussed in the next section.

But how do customers get this data into Amazon S3 in the first place?

Customers have several options for this. In the AV space, the customer uses the **Snow** family of devices (**SnowBall**, **SnowCone**, and **SnowMobile**) to transfer up to PB of data to Amazon S3. Customers with on-prem systems can also use **Amazon Outposts** to temporarily host and process this data with APIs similar to the cloud and also use **Amazon Direct Connect** to securely transfer data to Amazon S3 using a dedicated network connection. Several use cases highlighted here are from public references of companies using AWS for actual AV development (see the *References* section for more information).

Customers can also use **AWS IoT FleetWise**, a service that, at the time of writing, is still under preview, to easily collect and transfer data from vehicles to the cloud in near real time. With IoT FleetWise, customers first model the vehicle sensors using the FleetWise designer, then they install the **IoT FleetWise Edge Agent** onto the compatible edge devices that are running on the vehicle, define data schemas and conditions to collect the data, and stream this data to **Amazon Timestream** or Amazon S3.

The data collected can be raw or processed sensor data, image, audio, video, RADAR, or LiDAR data. Once this data is on Amazon S3, it can be processed, analyzed, and labeled before using it in downstream tasks. Several customers, such as TuSimple, use Amazon **Elastic Compute Cloud** (**EC2**) to do various AV-related processing tasks. Customers trying to optimize the cost of processing with this scale of data use EC2 Spot Instances extensively. In 2020, Lyft Level 5 reported that more than 75% of their computing fleet was on EC2 Spot Instances to reduce the cost of operation (see the *References* section for a link to the use case).

For image- and video-based preprocessing workloads, several pre-trained ML models can be used but require access to GPU-based instances. Toyota Research Institute, for example, extensively uses P3 and P4 instances for highly scalable and performant cloud applications. Companies such as Momenta also use Amazon EMR to create analytics services such as safe driving assistance decision services.

For labeling data (primarily image, video, and 3D LiDAR data), customers use **Amazon SageMaker Ground Truth** on AWS, which provides specialized templates for these labeling use cases, access to private, vendor, or public labeling worker pools, and several assistive labeling capabilities to speed up these time-intensive tasks and reduce costs.

Customers also use pipelines to orchestrate end-to-end preprocessing, training, or post-processing workflows. A service that can help create, manage, and run these pipelines at scale is Amazon **Managed Workflows for Apache Airflow** or **MWAA**. MWAA is a managed orchestration service for Apache Airflow that makes it very simple to set up and operate end-to-end data pipelines in the cloud at scale. Alternatives to using MWAA include AWS services such as **Step Functions** and **Amazon SageMaker Pipelines**.

Model training, simulation, model compilation, verification, and validation workflows can make use of **Amazon EC2**, or one of the following managed services – **Amazon SageMaker**, **Amazon Elastic Kubernetes Service (EKS)**, **Amazon Elastic Container Service (ECS)**, and/or **AWS Batch**.

In the next section, we will discuss a reference architecture for AV development on AWS that brings together many of these services.

Designing an architecture for AV systems

In this section, we will be discussing a reference architecture published by AWS called the *Autonomous Driving Data Lake Reference Architecture*, a link to which can be found in the *References* section.

The complete architecture is replicated in *Figure 13.3*:

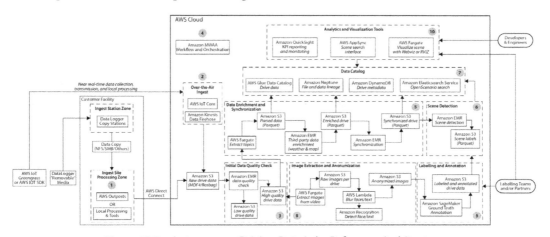

Figure 13.3 – Autonomous Driving Data Lake Reference Architecture

In this section, we will zoom into parts of this architecture to discuss it in further detail. Let's start with data ingestion:

- *Figure 13.4* shows how cars may be installed with a data logger or some removable storage media that stores data from sensors. Custom hardware or AWS Outposts can be used to process data that is stored from one or more trips. For near real time, **AWS IoT core** can be used along with **Amazon Kinesis Firehose** to deliver data to Amazon S3. Customers can also use Amazon Direct Connect, as mentioned earlier in this chapter, for secure and fast data transfer to Amazon S3.

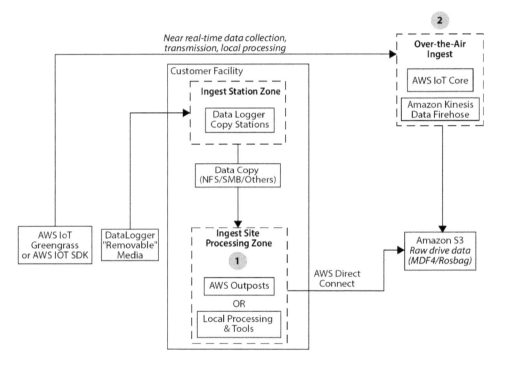

Figure 13.4 – Data ingestion – steps 1 and 2

- *Figure 13.5* shows how **Amazon EMR** can be used to filter incoming raw data. For example, the quality of data can be assessed using custom PySpark code and dropped into different buckets.

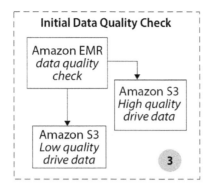

Figure 13.5 – Initial data processing – step 3

- *Figure 13.6* shows *step 5* and *step 6*, where data is further cleaned and enriched (for example, with location-specific or weather data). Given the large amount of data being collected, another processing step in Amazon EMR can be used to identify interesting scenes for downstream steps such as ML.

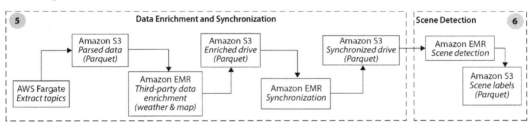

Figure 13.6 – Data enrichment – steps 5 and 6

- *Figure 13.7* shows how Amazon MWAA can be used to orchestrate these end-to-end data processing workflows (*step 4*). Data generated in many intermediate steps can be cataloged in **AWS Glue Data Catalog**, and the lineage of this data can be stored as a graph in **Amazon Neptune** (*step 7*).

- Finally, data can be preprocessed for visualization tools in Fargate tasks, with **AWS AppSync** and **Amazon QuickSight** providing visualization and KPI reporting capabilities.

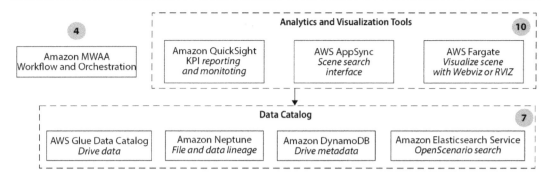

Figure 13.7 – Workflow Orchestration, Data Catalog and Visualization Tools – steps 4, 7, and 10, respectively

- *Figure 13.8* shows how **Amazon Fargate** tasks can be used to extract images or sequences of images from videos, with **AWS Lambda** functions used with pre-trained models or with the help of **Amazon Rekognition** to blur and anonymize parts of images such as faces or license plates. For AV customers, further pre-labeling can be done where pre-trained models available in open source can be used to identify pedestrians, cars, trucks, road signs, and so on.

This helps the labeling process go faster since most of the labeling effort is only to adjust existing labels, compared to creating all labels (such as bounding boxes) from scratch. This high-quality labeled data is the most important step when creating ML-powered ADAS and AV systems, which can include multiple models. More details on ML for AV systems will be discussed in the next section.

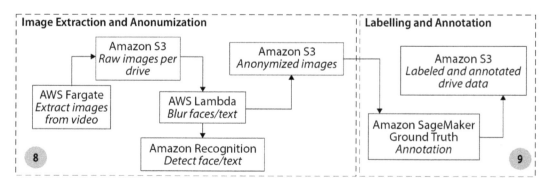

Figure 13.8 – Labeling and ML on labeled data – steps 8 and 9

In the next section, we will discuss how ML is applied to AV systems and use cases.

ML applied to AV systems

Developing highly sophisticated **Deep Neural Networks** (**DNNs**) with the ability to safely operate an AV is a highly complex technical challenge. Practitioners require PB of real-world sensor data, hundreds of thousands, if not millions, of **virtual Central Processing Unit** (**vCPU**) hours, and thousands of accelerator chips or **Graphics Processing Unit** (**GPU**) hours to train these DNNs (also called *models* or *algorithms*). The end goal is to ensure these models can operate a vehicle autonomously safer than a human driver.

In this section, we'll talk about what is involved in developing models relevant to end-to-end AV/ADAS development workflows on AWS.

Model development

AVs typically operate through five key processes, each of which may involve ML to various degrees:

- Localization and mapping
- Perception
- Prediction
- Planning
- Control

Each of the steps also requires different supporting data and infrastructure to efficiently produce a functional model or models. For example, while the perception stack is built on top of large computer vision models requiring distributed compute infrastructure to support DL training, the control step consumes a mix of general-purpose GPU and large memory GPU cards optimized for DL, in an online or offline **Reinforcement Learning** (**RL**) workflow.

In the next section, we will explore some of the challenges that a cloud-based ML environment can overcome for successful AV development by leveraging AWS.

Challenges

There are three main challenges in building DL-based AV models:

- Feeding TB or more of training data to ML frameworks running on large-scale, high-performance computing infrastructure
- Elasticity to linearly scale compute infrastructure to thousands of accelerators leveraging high bandwidth networking
- Orchestration of training the ML frameworks

Large amounts of data also means a large number of resources needed for labeling, so let us discuss this challenge next.

Labeling large amounts of data

Vehicle sensor and simulation data contain streams of images and videos, point clouds from RADAR and LiDAR, and time series data from inertia measuring sensors, GPS, and the vehicle **Controller Area Network (CAN)** bus. In a given test drive of 8 hours, an AV may collect more than 40 TB of data across these sensors. Depending on the company, a test fleet can range anywhere from a handful of vehicles to nearly 1,000. At such a scale, AV data lakes grow at PB annually. Altogether, this data is indexed by time and stored as scenes or scenarios leveraged for model training. For the purposes of building models and agents that ultimately drive the vehicle, the data needs to get processed and labeled.

The first challenge is to make these labeled datasets available to the ML framework during training as fast as it can process a data batch. The amount of data required to train one model in one specific task alone can be in excess of hundreds of TB, making pre-fetching and loading data into memory unfeasible. The combination of ML framework and compute hardware accelerator dictates the speed at which a batch of such data gets read from the source for a specific model task.

Today, you can use several built-in labeling templates on Amazon SageMaker Ground Truth for labeling images, video, and LiDAR data. For a description of AWS services that can be used to preprocess and label high-resolution video files recorded, visit the linked blog in the *References* section. Specifically for LiDAR use cases, you can use data that has both LiDAR as well as image data captured in sync with the LiDAR using multiple onboard cameras. Amazon SageMaker Ground Truth can synchronize a frame containing 3D point cloud data with up to eight camera sources. Once the raw data manifest is ready, you can use SageMaker Ground Truth for 3D object detection, object tracking, and semantic segmentation of 3D point clouds. As with standard SageMaker Ground Truth labeling jobs, you can use a fully private workforce or a trusted vendor to complete your labeling tasks. *Figure 13.9* is an example of a car being labeled using a 3D bounding box along with three projected side views and images from cameras corresponding to that timestamp:

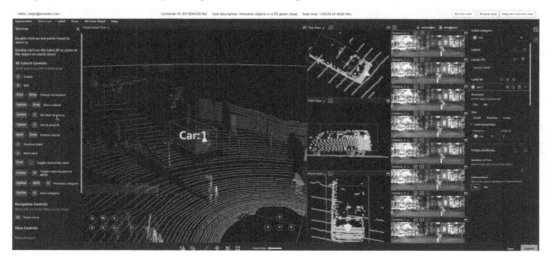

Figure 13.9 – Amazon SageMaker Ground Truth labeling jobs
with LiDAR and camera data for AV workloads

For more information on using Ground Truth to label 3D point cloud data, see the link in the *References* section.

Training with large amounts of data

Typically, DL tasks in the context of AV can be related to *perception* (finding information about the environment or obstacles) and *localization* (finding your position in the world with high accuracy). Several state-of-the-art DL models are being developed for detecting other vehicles, roads, pedestrians, signs, and objects and also to describe the position of these 2D objects around the vehicle in the 3D world. The *KITTI Benchmark* is often used to test new algorithms and approaches to looking at use cases related to autonomous vehicles, such as semantic segmentation and object detection. Access to the KITTI dataset (and other similar datasets such as the *Audi A2D2 Autonomous driving dataset*) can be found on the Registry of Open Data on AWS.

Training large semantic segmentation and object detection models on large open source datasets, such as **Mask R-CNN** on the **Common Objects in Context** (**COCO**) dataset, can achieve throughputs of 60 images per second – approximately 35 MB/s – on a single multi-GPU instance. For simpler architectures, training throughput can reach thousands of images per second due to the smaller scale of the network being trained on more straightforward tasks. That is true in the case of image classification using *ResNet* models that make up the backbone of larger models such as *Mask R-CNN*. More recently, models such as *DeepManta* have been used to obtain high scores on other related tasks such as vehicle pose estimation – links to these methods and papers can be found in the *References* section.

At lower rates of data transfer, the training job can retrieve data objects directly from Amazon S3. Amazon S3 is a foundational data service for AV development on AWS, as discussed in this blog post on *Building an Autonomous Driving Data Lake* (see the *References* section).

Some data loaders provide connectivity directly to Amazon S3, such as TensorFlow, for TFRecord-style datasets. However, it requires optimizing the size of each data file as well as the number of worker processes to maximize Amazon S3 throughput, which can make the data loading pipeline complex and less scalable. It is possible to achieve hundreds of GB/s total throughputs reading directly from S3 when horizontally scaling the data reader, but there is a compromise on CPU utilization for the training process itself. This blog (https://aws.amazon.com/blogs/machine-learning/optimizing-i-o-for-gpu-performance-tuning-of-deep-learning-training-in-amazon-sagemaker/) explains how to optimize I/O for GPU performance. Mobileye explained their use of TFRecord datasets and Pipe mode on Amazon SageMaker for training their large DNNs resulting in faster training and a 10x improvement in development time in their reinvent video , which is included in the *References* section.

For a more straightforward architecture, which leverages the ML framework's native support of *POSIX-compliant* file system interface, AWS offers **FSx for Lustre** (more information about FSx and POSIX compliance is linked to in the *References* section). FSx for Lustre is a high-throughput, low-latency distributed file system that can be provisioned from existing S3 data, making the whole dataset available to the DNN training workers as files. These files can be iterated over using any of the major ML framework data readers, such as PyTorch dataset or DataLoader.

FSx for Lustre can scale its baseline aggregate bandwidth to 200 GB/s for a 1 PB training dataset, with burst speeds of 1.3 GB/s per TiB of training data. The larger the provisioned FSx for Lustre deployment, the higher the aggregate bandwidth, enabling a PB-scale network fabric. FSx for Lustre is hydrated with a subset of the data from the Autonomous Driving data lake and synchronized back using data repository tasks in case model artifacts or data transformations are generated and recorded during training. For a real-world example of how the Amazon ML Solutions Lab helped Hyundai train a model using SageMaker's distributed training library and FSx for Lustre 10x faster with only 5x the number of instances, take a look at the link in the *References* section to this use case.

The need for a PB-scale data repository also comes from the need to scale the number of compute workers processing this data. At the 60 images per second rate, a single worker would take more than 6.5 hours to train over the 118,000 images in the *COCO* dataset, considering a dozen epochs to achieve reasonable accuracy. Scaling the number of images per training iteration is key for achieving reasonable training times. Even more so given the experimental and iterative nature of building DL based models, requiring multiple training runs for a single model to be built. Large-scale training generally translates to high costs of running training experiments. Amazon SageMaker provides cost-saving features for both training as well as deployment.

Alex Bain, Lead for ML Systems at Lyft Level 5, said: "*By using Amazon SageMaker distributed training, we reduced our model training time from days to a couple of hours. By running our ML workloads on AWS, we streamlined our development cycles and reduced costs, ultimately accelerating our mission to deliver self-driving capabilities to our customers.*"

Visit the following blog post for more examples of use cases around cost savings: `https://aws.amazon.com/blogs/aws/amazon-sagemaker-leads-way-in-machine-learning/`.

Amazon SageMaker provides connections to common file systems that store data, such as Amazon S3, EFS, and FSx for Lustre. When running long training jobs, the choice of an appropriate storage service can speed up training times overall. If training data is already in EFS, it is common to continue preprocessing data on EFS and training the model by pointing SageMaker to EFS. When data is in Amazon S3, customers can decide to use this data directly from S3 to make use of features on SageMaker such as Fast File mode, Pipe mode, data shuffling, or sharding by S3 key for distributed training (more information about these modes is included in the *References* section). Customers can also use FSx for Lustre since it automatically makes data available to SageMaker training instances and avoids any repetitive copying of data. When multiple epochs use slightly different subsets of data that fit into instance memory, or in the case of distributed training, FSx for Lustre provides extremely fast and consistent access to datasets by mounting a volume with the data accessible to your training code.

Scaling

With distributed training strategies, many compute nodes within a cluster read batches of data, train over them, and synchronize the model parameters as the training goes on. The unit of compute for these clusters is not the individual compute instance, sometimes called a *node*, but the individual GPUs. This is because the DNNs require hardware acceleration for training. So, distribution occurs within and across multi-GPU compute instances.

Amazon's EC2 service provides the broadest compute platform in the cloud with 17 distinct compute instance families. Each family is designed for a few specific workloads and consists of a given ratio of vCPU, GPU (for certain instances), memory, storage, and networking. For full, end-to-end AV development, companies largely rely on the C, M, R, G, and P instance families.

For ML model training, companies leverage the **Deep Learning Amazon Machine Images** (**DLAMI**) to launch NVIDIA GPU-based EC2 instances in the *P family*. Each EC2 P family instance generation integrates the latest NVIDIA technology, including the p2 instances (Tesla K80) and the p3 instances (Volta V100), and the recently released p4d (with Ampere A100 GPUs).

AWS and NVIDIA continue to collaborate on achieving state-of-the-art model training times for data-parallel and model-parallel training (see the blog link in the *References* section). SageMaker distributed training includes libraries for distributed data-parallel and model-parallel modes of training. These libraries for data and model-parallel training extend SageMaker's training capabilities, so you can train large-scale models with small code changes to your training scripts. For readers interested in this topic, the following video from the AWS Deep Engines team is on an AWS library that is useful for both data- and model-parallel training: `https://youtu.be/nz1EwsS5OiA`.

At the single GPU level, optimizing the memory consumption helps increase the throughput. For model training on data that can be batched, this means increasing the number of images per iteration before running out of GPU memory. Therefore, the higher the GPU memory, the greater the achievable training throughput, which favors large memory GPU nodes.

Across GPUs, fast communication within and between instances enables faster synchronization of gradients during training. Networking is, therefore, a key aspect of scalability in enhancing the speed of each iteration step. This type of infrastructure is analogous to a tightly coupled HPC infrastructure.

AWS offers EC2 instances that support HPC and accelerated computing on the cloud. AWS has demonstrated the fastest training times for models such as Mask R-CNN and near linear scalability of large-scale GPU instances using the EC2 **p3dn.24xlarge** instance. This instance has 8 NVIDIA V100 GPUs, with 32 GB of memory each, and can make use of the **AWS Elastic Fabric Adapter** (**EFA**) network interface. EFA is a custom-built OS bypass hardware interface that enhances the performance of inter-instance communication, achieving 100 gigabits per second bandwidth per card and natively integrating with communication libraries such as MPI and **NVIDIA Collective Communication Library** (**NCCL**) used on ML applications.

AWS introduced the latest generation (in 2020) of NVIDIA GPU hardware **General Availability** (**GA**) with EC2 p4d instances. This instance takes ML training in the cloud to the next level and includes 8 NVIDIA A100 GPUs with 40 GB of memory per GPU and an improved networking stack. Instead of a single network interface card, the p4d instance has 4 EFA cards for a total of 400 gigabits per second bandwidth. Within the instance, the p4d family also increases GPU-to-GPU communication bandwidth with an NVlink mesh topology for ML frameworks using NCCL. The new p4d instance design provides up to 3.8x the training throughput compared to the **p3dn.24xlarge** on backbone models for the major computer vision tasks, such as semantic segmentation, used in the data labeling phase of the AV development process. For more information on the p4d design and benchmark results, refer to this deep dive blog.

During the December 2020 AWS re:Invent conference, AWS announced plans to make Intel Habana Gaudi accelerators and an in-house built training chip, which will offer more **Terraflops (TFLOPS)** than any compute instance in the cloud, available. In 2020, AWS collaborated with NVIDIA to bring down training times of Mask R-CNN on the cloud to 6 minutes and 45 seconds on PyTorch and 6 minutes and 12 seconds with TensorFlow (see the link in the *References* section). For more information on data-parallel and model-parallel training with minimal code changes on Amazon SageMaker, see the documentation on SageMaker distributed training linked to in the *References* section.

Orchestration

The final challenge in training AV DNNs is managing and orchestrating the tightly coupled HPC infrastructure at scale. AWS provides a suite of services and solutions for HPC that you can leverage to build and manage a large-scale DNN training cluster for AV including tools such as Amazon EKS, Amazon ECS, AWS Batch, and AWS Parallel Cluster. These topics have been discussed in the preceding chapters in detail and will not be repeated here.

The infrastructure management challenge also includes the ability to integrate upstream and downstream tasks from the AV stack throughout the development of the models. As an example, the validation of a perception model stack may include online driving simulations. When integrating a simulation environment into a model-building deployment, the requirements for distributed compute change from tightly coupled high-performance computing to highly parallelized, embarrassingly parallel batch simulations and client-server architectures. Efficiently integrating services becomes critically important to bringing AV systems development from a research exercise to a scalable, production-ready pipeline.

These options give you the flexibility to build a scalable and diverse ML platform for your data scientist team with the quickest velocity and most robust infrastructure in the world. Regardless of whether you use a managed platform for ML such as Amazon SageMaker, or manage your own platform on Kubernetes, ML deployment and orchestration specifically for AV needs to have the following:

- Continuous training and re-training functionality
- Continuous deployment
- Continuous monitoring

More information about MLOps on AWS can be found at `https://aws.amazon.com/sagemaker/mlops/`.

For AV specifically, here is what a typical ML workflow spanning a few weeks may look like:

1. New or updated datasets.
2. Dataset curation and scene selection.
3. Pre-labeling and data curation, where pre-trained models are used to provide coarse quality labeled data to human labelers, and in some cases where sensitive information is obfuscated from image and video datasets.

4. Labeling and active learning (more information at `https://aws.amazon.com/sagemaker/data-labeling/`).

5. Distributed training for various tasks (see *Chapter 5, Data Analysis*).

6. Model testing:

   - Software-in-the-loop testing

   - Hardware-in-the-loop testing

   - On-road testing

7. Dataset collection, and go back to *step 1*.

For more information about these steps, you may be interested in reading about how an actual customer of AWS, Aurora, achieves self-driving capabilities using their Aurora Driver platform, at `https://www.cnet.com/roadshow/news/aurora-drive-aws-autonomous-vehicle-development/` and `https://www.youtube.com/watch?v=WAELZY_TJ04`.

Chapter 6, Distributed Training of Machine Learning Models, discussed distributed training, and *Chapter 8, Optimizing and Managing Machine Learning Models for Edge Deployment*, discussed model deployment at the edge; these are both very relevant to the topics listed in the previous steps (*steps 5–9*). Within model testing, software-in-the-loop testing can be done using tools on AWS, and this will be discussed in the next section using a hands-on example that you can follow along on your AWS account.

Software-in-the-loop (SITL) simulation

In this section, we will be discussing one very specific type of **Software-in-the-Loop** (SITL) testing that is useful for AV customers. Note that this is not the only type of simulation that is being run by AV customers around the world today. Some may involve perception tasks, planning or mapping tasks, and also end-to-end software tasks before moving on to **Hardware-in-the-Loop** (HITL) or on-road testing.

In this section, we will walk through how you can set up a high-fidelity simulation of a driving environment and even test out some DL models for AV within the simulation environment! To do this, you have to follow two high-level steps:

1. Create a container with your simulation environment.

2. Use RoboMaker to run your simulation.

Once you set this up, you can interactively work with your simulation, use the environment as part of a RL loop, or even generate synthetic data for your future ML experiments.

Before we walk through the steps, here are some basics:

- *What is AWS RoboMaker?* **RoboMaker** is a cloud-based simulation service where you can run your simulations without managing any infrastructure. More information can be found at `https://aws.amazon.com/robomaker/`. RoboMaker provides GPU-based compute for high-fidelity simulators discussed in the following point.

- Specific to AV and manufacturing-related simulations, RoboMaker lets you use simulators such as CARLA (which we will be using in this section), AirSim, Aslan, Summit, DeepDrive or *Ignition*, Drake or NVIDIA Isaac Sim, and even custom simulations that you create using high-fidelity rendering engines such as Unity or Unreal Engine. Links to these simulators and tools can be found in the *References* section.

- **CARLA** is an open source project that is commonly used in AV studies for simulating vehicles in environments and testing out DL or RL models for AV. CARLA exposes an API that lets users control all aspects of the simulation, such as driving, environment, traffic, and pedestrians, and also lets users configure sensors on vehicles, such as LiDARs, cameras, and GPS.

Great, let us now get started with the steps required to run CARLA on RoboMaker!

Step 1 – build and push the CARLA container to Amazon ECR

To build our custom simulation container, we need two files:

- A Dockerfile
- A shell script to build and push the customer's Docker container to ECR (you can use your own pipeline to do this step as well)

Here, we will be building the container on a SageMaker notebook instance, but you may use the same script from your local laptop or an EC2 instance provided you have the right permissions set up. To demonstrate this step, we will assume that you're aware of SageMaker notebooks.

Here is what the Dockerfile looks like for CARLA when viewed from a notebook instance. The `writefile` command writes the following code into a new file named `Dockerfile`:

```
%%writefile Dockerfile


FROM carlasim/carla:0.9.11
USER root


# see https://github.com/NVIDIA/nvidia-docker/issues/1632
RUN rm /etc/apt/sources.list.d/cuda.list
RUN rm /etc/apt/sources.list.d/nvidia-ml.list
```

```
# install dependencies
RUN apt update && \
      apt install -y python3-pip \
      libjpeg-dev \
      libtiff5-dev \
      libomp-dev \
      fontconfig

# fix ALSA errors
RUN echo pcm.!default { type plug slave.pcm "null" } >> /etc/
asound.conf

# install NICE DCV (for RoboMaker)
RUN apt update -y && apt upgrade -y && apt install -y wget pgp
RUN wget https://d1uj6qtbmh3dt5.cloudfront.net/NICE-GPG-KEY
RUN gpg --import NICE-GPG-KEY
RUN wget https://d1uj6qtbmh3dt5.cloudfront.net/2021.1/Servers/
nice-dcv-2021.1-10598-ubuntu1804-x86_64.tgz
RUN tar -xvzf nice-dcv-2021.1-10598-ubuntu1804-x86_64.tgz
RUN apt update && apt install -y ./nice-dcv-2021.1-10598-
ubuntu1804-x86_64/nice-dcv-gl_2021.1.937-1_amd64.ubuntu1804.deb
\
                              ./nice-dcv-2021.1-10598-
ubuntu1804-x86_64/nice-dcv-gltest_2021.1.275-1_amd64.
ubuntu1804.deb

# install opengl
RUN apt update && apt install -y libglfw3 libglfw3-dev

# install xterm
RUN apt update && apt install -y xterm

# run as user carla
USER carla

# install example dependencies
```

```
RUN python3 -m pip install -U pip
RUN cd ~/PythonAPI/examples && python3 -m pip install -r
requirements.txt

# set path to carla python API
ENV PYTHONPATH=/home/carla/PythonAPI/carla/dist/carla-0.9.11-
py3.7-linux-x86_64.egg

ENTRYPOINT ["/bin/bash", "-c"]
```

As you can see, we start off with the CARLA base image and install some additional dependencies:

```
FROM carlasim/carla:0.9.11
```

The second file that is required is a script to build and push the container. First, we define the arguments (the name of the container) and some inputs to the script, such as region, account, and the full name of the Docker container to be built:

```
%%sh

# The name of our algorithm
algorithm_name=$1

account=$(aws sts get-caller-identity --query Account --output
text)

# Get the region defined in the current configuration (default
to us-west-2 if none defined)
region=$(aws configure get region)

fullname="${account}.dkr.ecr.${region}.amazonaws.
com/${algorithm_name}:latest"
```

Then, we create the repository and log into it:

```
# If the repository doesn't exist in ECR, create it.
aws ecr describe-repositories --repository-names "${algorithm_
name}" > /dev/null 2>&1
```

```
if [ $? -ne 0 ]
then
    aws ecr create-repository --repository-name "${algorithm_
name}" > /dev/null
fi


# Get the login command from ECR and execute it directly
aws ecr get-login-password --region ${region}|docker login
--username AWS --password-stdin ${fullname}
```

Finally, we build and push the container to ECR:

```
# Build the docker image locally with the image name and then
push it to ECR
# with the full name.

docker build  -t ${algorithm_name} .
docker tag ${algorithm_name} ${fullname}


docker push ${fullname}


echo ${fullname}
```

This shell script takes the name of your container as an argument, builds the container based on a local Dockerfile, and pushes the container to a repository in ECR. Once you have these two scripts on a notebook instance or anywhere with Docker installed, you can run the following:

```
./build_and_push.sh carlasim
```

This will output the location of the container you just built, similar to the following:

```
<account_number>.dkr.ecr.<region>.amazonaws.com/carlasim:latest
```

Copy this output, as you will need it in *Step 2 – configure and run CARLA on RoboMaker*.

Step 2 – configure and run CARLA on RoboMaker

To configure and run CARLA simulations on RoboMaker, go through the following steps:

1. Navigate to the **AWS RoboMaker** console on AWS – `https://aws.amazon.com/robomaker/` (sign in to your AWS account if asked).

2. Next, look for **Simulation Jobs** on the left sidebar menu and click **Create simulation job**. You should now see a screen similar to *Figure 13.10*. Leave the default simulation job duration as is (**8 hours**), and create a new role (see arrow) called `carsimrole`:

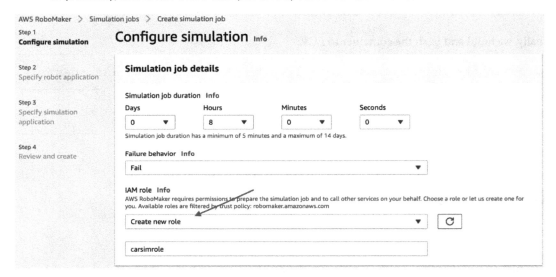

Figure 13.10 – Create a simulation job on AWS RoboMaker

3. On the same screen, scroll down to the compute options and make sure you select CPU and GPU, with sliders corresponding to **Simulation Unit (SU) limit** and **GPU Unit (GPU) limit** all the way to the right (maximum), as shown in *Figure 13.11*. Also, browse to a known location to save the outputs of your simulation job on Amazon S3. In *Figure 13.11*, the location selected shows `s3://carlasim-bucket/output`; make sure you browse to a bucket that you have access to and not the one shown in this example:

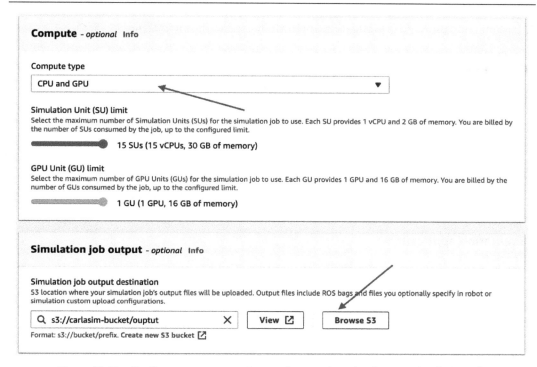

Figure 13.11 – Configure compute options and output location for your simulation job

4. On the **Specify robot application** page, select **None** for robot application (see *Figure 13.12*):

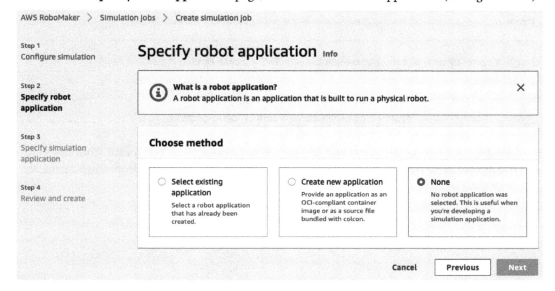

Figure 13.12 – Select None for robot application

5. Move on to the **Specify Simulation application** page, as shown in *Figure 13.13*:

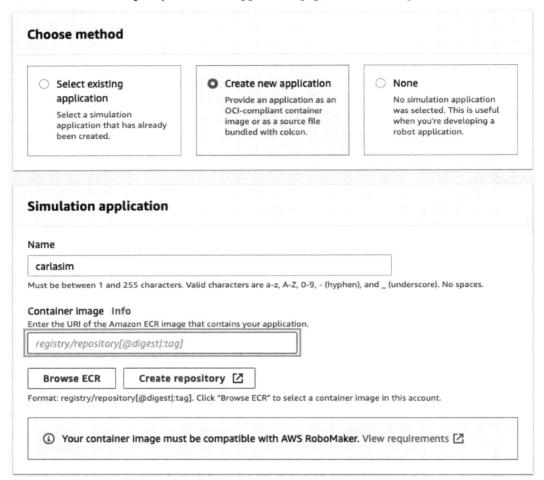

Figure 13.13 – Create a new simulation application

6. Select **Create new application,** and add the ECR repository link you copied from *step 1*. If you already have a simulation application, you can choose it from the drop-down menu, as shown in *Figure 13.14*:

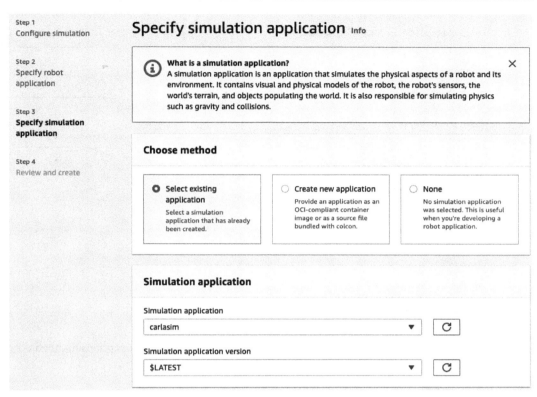

Figure 13.14 – Configure Simulation application by selecting it from the drop-down menu

7. Scroll down to the **Simulation application configuration** section, enter the following as your launch command, as shown in *Figure 13.15*, and remember to check the option for running with a streaming session:

```
./CarlaUE4.sh -opengl
```

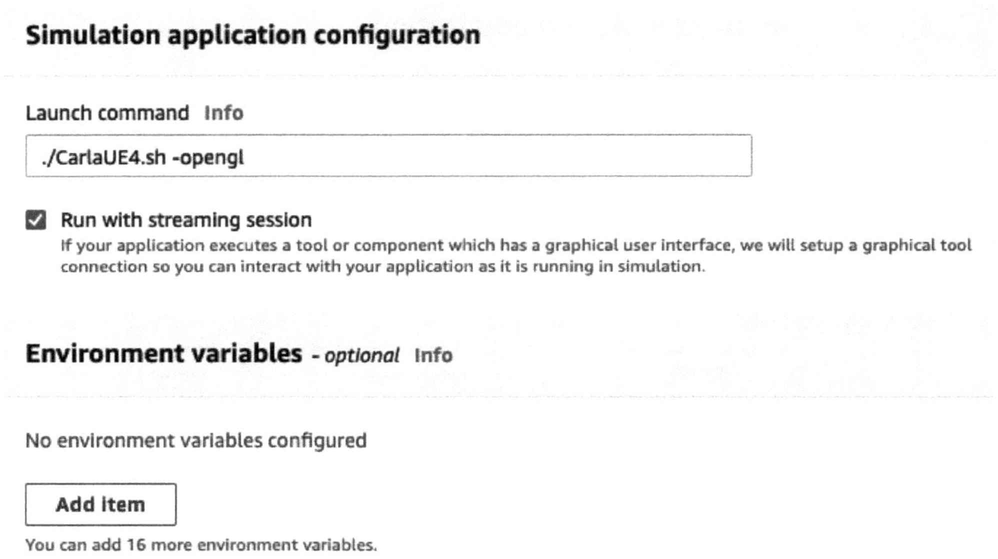

Simulation application configuration

Launch command Info

```
./CarlaUE4.sh -opengl
```

☑ **Run with streaming session**
If your application executes a tool or component which has a graphical user interface, we will setup a graphical tool connection so you can interact with your application as it is running in simulation.

Environment variables - *optional* Info

No environment variables configured

Add item

You can add 16 more environment variables.

Figure 13.15 – Enter the launch command in the Simulation application configuration section

8.   In the **Simulation application tools** section, create two terminals with the options highlighted in *Figure 13.16*:

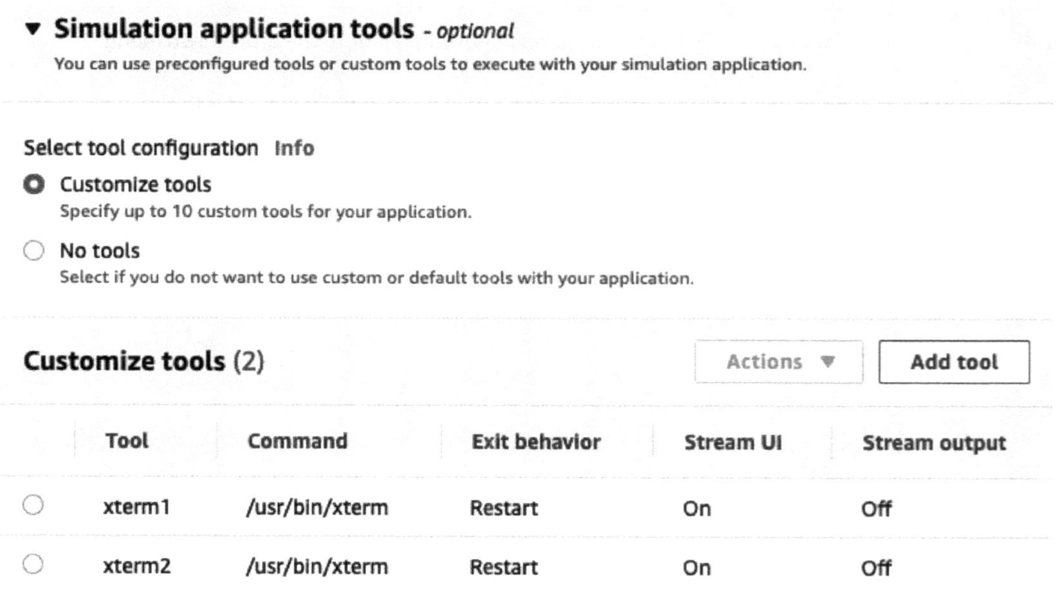

▼ Simulation application tools - *optional*

You can use preconfigured tools or custom tools to execute with your simulation application.

Select tool configuration Info

◉ **Customize tools**
Specify up to 10 custom tools for your application.

○ **No tools**
Select if you do not want to use custom or default tools with your application.

Customize tools (2)   Actions ▼   Add tool

	Tool	Command	Exit behavior	Stream UI	Stream output
○	xterm1	/usr/bin/xterm	Restart	On	Off
○	xterm2	/usr/bin/xterm	Restart	On	Off

Figure 13.16 – Add two custom tools for terminal access

9. Finally, click **Next** to go to the summary screen, and then click **Create**. This process of creating your simulation environment will take a few minutes. Once it is created, you should see your simulation job was created, as seen in *Figure 13.17*:

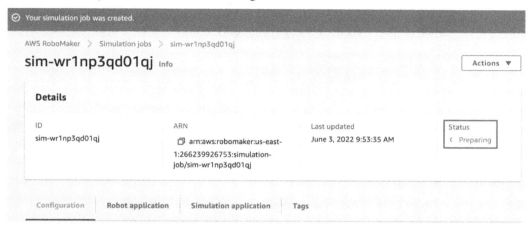

Figure 13.17 – Wait for the simulation job to be created

10. Once the simulation is created and the **Status** field says **Running**, click the **Connect** button in the **Simulation application** section, as shown in *Figure 13.18*. You can also access the terminal for running scripts or monitoring the simulation environment.

Figure 13.18 – Connect to the main simulation application or the terminals created

11. Click **Connect** on one of the **xterm** applications, **xterm1** or **xterm2**, and explore the file system; you should see Python examples inside the `PythonAPI` folder, as shown in *Figure 13.19*.

```
xterm (on efdf81afc8d8)

$ ls
CHANGELOG         Manifest_DebugFiles_Linux.txt
CarlaUE4          NICE-GPG-KEY
CarlaUE4.sh       Plugins
Co-Simulation     PythonAPI
Dockerfile        README
Engine            Tools
HDMaps            VERSION
Import            nice-dcv-2021.1-10598-ubuntu1804-x86_64
ImportAssets.sh   nice-dcv-2021.1-10598-ubuntu1804-x86_64.tgz
LICENSE
$ cd PythonAPI
$ ls
carla  examples  python_api.md  util
$ ▮
```

Figure 13.19 – CARLA Python examples inside the simulation job

Visit the getting started guide to understand the Python API as well as included examples (`https://carla.readthedocs.io/en/0.9.2/getting_started/`). Some examples are provided in *Figure 13.20* to *Figure 13.22*.

- *Figure 13.20* shows a sample application where you can manually drive a Tesla Cybertruck in the CARLA simulation with 264 other vehicles:

Figure 13.20 – Driving a Tesla Cybertruck manually in a CARLA simulation

- *Figure 13.21* shows a sample application that uses a Python program through one of the terminal applications we created to spawn traffic:

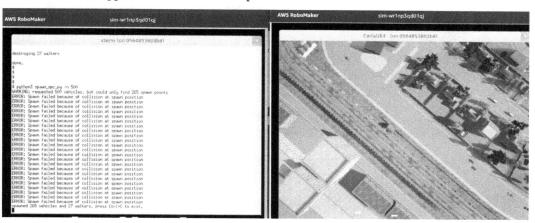

Figure 13.21 – Spawning traffic in the CARLA simulation world using Python code

- *Figure 13.22* shows an application that simulates LiDAR data around a vehicle in the simulation:

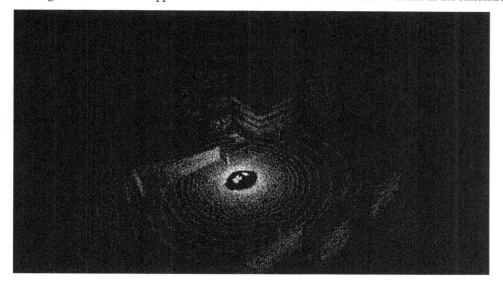

Figure 13.22 – Simulated LiDAR data around the car

As a next step, read how you can use RL models to control your car for self-driving use cases in this tutorial: https://carla.readthedocs.io/en/latest/tuto_G_rllib_integration/

Let's summarize all that we've learned so far in this chapter.

Summary

In this chapter, we discussed AV and ADAS systems at a high level, along with a reference architecture to build some of these systems on AWS. We also discussed the three main challenges practitioners face when training AV-related ML models in the cloud: feeding TB or more of training data to ML frameworks running on a large-scale, high-performance computing infrastructure, elasticity to linearly scale compute infrastructure to thousands of accelerators leveraging high bandwidth networking, and orchestrating the ML framework training.

Lastly, we walked you through examples of how you can make use of tools on AWS to run SITL simulations for testing your ML models.

In the next chapter, we will focus on solving numerical optimization problems on AWS.

References

For more information about topics discussed in this chapter, visit the following links:

- *Autonomous Vehicle and ADAS development on AWS Part 1: Achieving Scale*: `https://aws.amazon.com/blogs/industries/autonomous-vehicle-and-adas-development-on-aws-part-1-achieving-scale/`
- *Building an Autonomous Driving and ADAS Data Lake on AWS*: `https://aws.amazon.com/blogs/architecture/field-notes-building-an-autonomous-driving-and-adas-data-lake-on-aws/`
- *Implementing Hardware-in-the-Loop for Autonomous Driving Development on AWS*: `https://aws.amazon.com/blogs/architecture/field-notes-implementing-hardware-in-the-loop-for-autonomous-driving-development-on-aws/`
- *Advanced Driver Assistance Systems (ADAS)*: `https://www.gartner.com/en/information-technology/glossary/advanced-driver-assistance-systems-adass`
- *CARLA Documentation*: `https://carla.readthedocs.io/en/0.9.2/getting_started/`
- *Highly Automated and Autonomous Vehicle Development with Amazon Web Services*: `https://pages.awscloud.com/rs/112-TZM-766/images/Autonomous_Vehicle_Development_with_AWS.pdf`)
- *AWS IoT FleetWise*: `https://aws.amazon.com/iot-fleetwise/`
- *Lyft Increases Simulation Capacity, Lowers Costs Using Amazon EC2 Spot Instances*: `https://aws.amazon.com/solutions/case-studies/Lyft-level-5-spot/`

- *Autonomous Driving Data Lake Reference Architecture*: https://d1.awsstatic.com/architecture-diagrams/ArchitectureDiagrams/autonomous-driving-data-lake-ra.pdf?did=wp_card&trk=wp_card

- *Automating Data Ingestion and Labeling for Autonomous Vehicle Development*: https://aws.amazon.com/blogs/architecture/field-notes-automating-data-ingestion-and-labeling-for-autonomous-vehicle-development/

- *Use Ground Truth to Label 3D Point Clouds*: https://docs.aws.amazon.com/sagemaker/latest/dg/sms-point-cloud.html

- *Mask RCNN paper*: https://arxiv.org/abs/1703.06870

- *COCO dataset*: https://cocodataset.org

- *KITTI dataset*: https://registry.opendata.aws/kitti/ and http://www.cvlibs.net/datasets/kitti/

- *A2D2 dataset*: https://registry.opendata.aws/aev-a2d2/

- *ResNet*: https://arxiv.org/abs/1512.03385

- *DeepManta*: https://arxiv.org/abs/1703.07570

- *Vehicle Pose Estimation on KITTI Cars Hard*: https://paperswithcode.com/sota/vehicle-pose-estimation-on-kitti-cars-hard.

- *Building an Autonomous Driving and ADAS Data Lake on AWS*: https://aws.amazon.com/blogs/architecture/field-notes-building-an-autonomous-driving-and-adas-data-lake-on-aws/

- *TFRecord dataset*: https://www.tensorflow.org/api_docs/python/tf/data/TFRecordDataset

- *Performance Design Patterns for Amazon S3*: https://docs.aws.amazon.com/AmazonS3/latest/userguide/optimizing-performance-design-patterns.html

- *Moving our Machine Learning to the Cloud Inspired Innovation*: https://www.mobileye.com/blog/moving-our-machine-learning-to-the-cloud-inspired-innovation/

- *Lustre User Guide*: https://docs.aws.amazon.com/fsx/latest/LustreGuide/what-is.html

- *Pytorch DataLoader*: https://pytorch.org/docs/stable/data.html

- *Exporting changes to the data repository*: https://docs.aws.amazon.com/fsx/latest/LustreGuide/export-changed-data-meta-dra.html

- *Hyundai reduces ML model training time for autonomous driving models using Amazon SageMaker*: https://aws.amazon.com/de/blogs/machine-learning/hyundai-reduces-training-time-for-autonomous-driving-models-using-amazon-sagemaker/

- *Access Training Data*: https://docs.aws.amazon.com/sagemaker/latest/dg/model-access-training-data.html

- *Amazon EC2 P4 Instances*: https://aws.amazon.com/ec2/instance-types/p4/

- *AWS and NVIDIA achieve the fastest training times for Mask R-CNN and T5-3B*: https://aws.amazon.com/blogs/machine-learning/aws-and-nvidia-achieve-the-fastest-training-times-for-mask-r-cnn-and-t5-3b/

- *NVLink and NVSwitch*: https://www.nvidia.com/en-us/data-center/nvlink/

- *NVIDIA Collective Communications Library (NCCL)*: https://developer.nvidia.com/nccl

- *AWS re:Invent*: https://reinvent.awsevents.com/

- *Amazon EC2 DL1 Instances*: https://aws.amazon.com/ec2/instance-types/habana-gaudi/

- *Distributed training libraries*: https://aws.amazon.com/sagemaker/distributed-training/

- *CARLA*: http://carla.org/

- *AirSim*: https://github.com/microsoft/AirSim

- *Project Aslan*: https://github.com/project-aslan/Aslan

- *SUMMIT Simulator*: https://github.com/AdaCompNUS/summit

- *Deepdrive*: https://github.com/deepdrive/deepdrive

- *Gazebo*: https://gazebosim.org/home

- *Drake*: https://drake.mit.edu/

- *NVIDIA Isaac Sim*: https://developer.nvidia.com/isaac-sim

- *Unity*: https://unity.com/

- *Unreal Engine*: https://www.unrealengine.com/

14

Numerical Optimization

In our daily lives, while running errands and doing chores, the human mind is always carrying out some form of optimization. For example, the mind might be optimizing the route to take for single or multiple destinations that we need to visit. It can also be optimizing the cost of items that we need to buy on a trip to a grocery store, or, for example, budgeting our income and expenses on a weekly or monthly basis. Another example is to try to optimize the amount of sleep so that our mind is fresh the following day to work on our projects. In short, we are optimizing multiple tasks and schedules every single day without even knowing or thinking about it. Similarly, nature also optimizes its processes. For example, the Earth goes around the Sun in an optimal path to keep a balance between the various gravitational forces.

Optimization also plays a big role in the technology industry. Several large-scale optimization problems are being solved by small and large corporations. For example, a courier delivering packages to our home follows the route and schedule assigned by an optimization problem that solved an equation (either numerically or analytically) under several constraints to come up with that optimal route. Similarly, stock trading is another example, where the action can be to sell, hold, or buy stocks of a particular company to maximize long-term or short-term gains.

In this chapter, we are going to discuss optimization in general while focusing more on numerical optimization, its examples, and use cases, along with its application in applied machine learning. The following topics will be covered in this chapter:

- Introduction to optimization
- Common numerical optimization algorithms
- Example use cases of large-scale numerical optimization problems
- Numerical optimization using high-performance compute on AWS
- Machine learning and numerical optimization

Introduction to optimization

As mentioned in the introduction to this chapter, optimization is an important tool for making decisions related to a large set of problems in our daily lives and various fields of science. There are various components to an optimization problem, as we are going to discuss in the following subsections.

Goal or objective function

The process of optimization starts with defining a goal or an objective, such as monetary gain, a route or path, a schedule, items, and so on. Selecting the goal or objective depends heavily on the problem domain, as well as the specific problem we are trying to solve. In addition to the objective function, we also need to know whether we are maximizing or minimizing the objective function. Again, this also depends on the specific problem domain, as well as the objective function. For an optimization problem with cost as the objective function, our goal will most likely be to minimize it, whereas if our objective function is revenue or profit, we would like to maximize it.

For our route optimization example, one organization might be focused on solving the problem to maximize the number of delivered items, while another organization might want to minimize fuel cost per delivery. So, even though the problem domain is the same for both problems, the objective is different. Many times, the objectives may be related or dependent on each other. For example, in the route optimization problem, the number of delivered items and the fuel cost per delivery seem to be dependent on each other. Trying to deliver the maximum number of items in a given amount of time also means that the route needs to be defined in such a way that the distance from one location to the next is short. This means the vehicle is going to travel short distances to deliver items and hence fuel cost per delivery will be less.

Variables

The objective function for any optimization problem is generally a function of several variables. By changing the values of these variables, the value of the objective function also changes. For example, for the route optimization problem, one of the variables can be the speed of the vehicle. If the vehicle speed is increased, the number of deliveries made by the vehicle will increase, thereby improving the objective function's value. Numerical optimization problems vary the values of these variables in an attempt to arrive at the optimal value of the optimization function. Mathematically, we can define the objective function, f, which maps some set of variables, X, to real space, $\mathbb{R}$:

$$f : X \rightarrow \mathbb{R}$$

Our objective or goal is to find the values, X^*, of the X variable that minimize or maximize (depending on the problem) our objective function, f. So, for the maximization case, the optimization problem can be written as follows:

$$f(X^*) = \max_X f(X)$$

For the minimization case, the optimization problem can be written as follows:

$$f(X^*) = \min_X f(X)$$

Constraints

In the route optimization problem discussed previously, there will be an upper limit on the speed of a vehicle based on the maximum speed allowed on a road. The vehicle should not exceed that speed limit. This will be a constraint on the optimization objective function. Any optimization problem will either have constraints (constrained optimization) or not (unconstrained optimization). Generally, a large-scale numerical optimization problem has several constraints. The goal of the optimization problem in the presence of constraints then becomes finding the best value of the objective function while satisfying all the constraints. A few example constraints are defined here:

- **Linear constraints**: The x_1 and x_2 variables are greater than or equal to zero and their sum is less than 100:

$$x_1 \geq 0,$$

$$x_2 \geq 0,$$

$$x_1 + x_2 < 100.$$

- **Non-linear constraints**: The square of the x_1 variable is greater than x_2:

$$x_1^2 > x_2$$

A real-world large-scale numerical optimization problem will generally have both linear as well as non-linear constraints.

Modeling an optimization problem

One of the hardest and most important tasks in any optimization problem is formulating or modeling the problem itself. This process involves identifying the variables, constraints, and objective function. Knowledge of the problem domain as well as a good understanding of the business problem that we are trying to solve are very important to have a good formulation of the optimization problem. Having a very simplistic formulation will not help us achieve good results with the problem, whereas having a too complicated formulation of the problem might result in giving us no solution or a bad solution while taking a very long time to numerically solve the problem, even on modern-day machines.

Optimization algorithm

After formulating the problem, the next step is to pick an optimization algorithm and then use a software tool to run it on the data containing our variables and constraints. No one algorithm solves all the optimization problems. Picking the right algorithm is a big factor in getting a good solution in a reasonable amount of time. Similarly, there are several open source as well as commercial tools with implementations of optimization algorithms. Depending on our budget and the resources available, we should pick the right software tool to solve the optimization problem. Once the algorithm has been executed and we have the results, the next step is to make sure that all the constraints, as well as optimality conditions, are satisfied. We can also carry out sensitivity analysis on the solution, if it is not an optimal solution, to improve upon it.

Local and global optima

The objective function that our optimization problem is attempting to solve usually has more than one optimum value. For example, if our optimization problem is a minimization problem and our objective function is convex, then it will have only one minimum value, called the **global minimum**, which can be found using methods based on calculus or well-known algorithms such as gradient descent, hill climbing, and so on.

Figure 14.1 shows the case of a convex objective function of one variable with a global minimum value, while *Figure 14.2* shows a convex objective function of two variables:

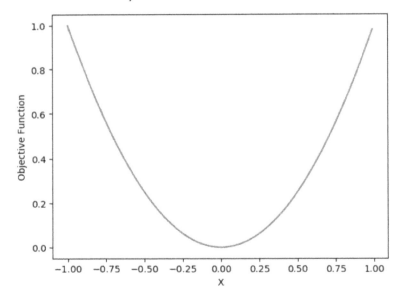

Figure 14.1 – A convex objective function of one variable with a global minimum

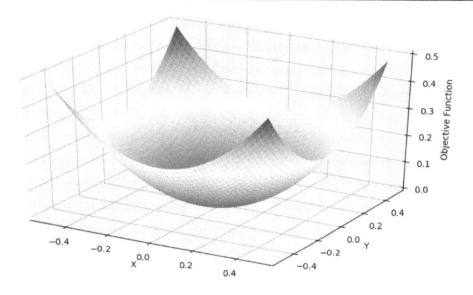

Figure 14.2 – A convex objective function of two variables with a global minimum

Most of the optimization problems that we encounter and process in our daily lives have non-convex objective functions. These objective functions have more than one optimum value, referred to as **local optima**. *Figure 14.3* and *Figure 14.4* show examples of objective functions with one variable with multiple local minima:

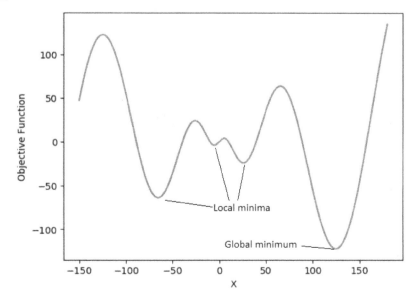

Figure 14.3 – Objective function with multiple local minima

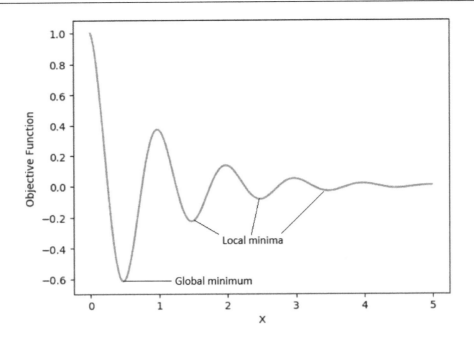

Figure 14.4 – Objective function with multiple local minima

Figure 14.5 and *Figure 14.6* show examples of objective functions with two variables with multiple local minima:

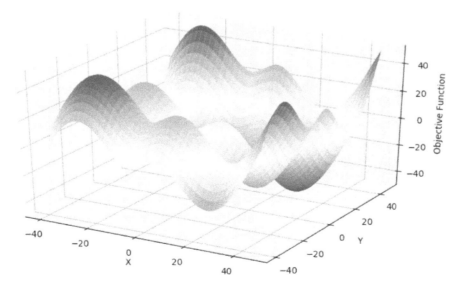

Figure 14.5 – Non-convex objective function showing multiple local minima

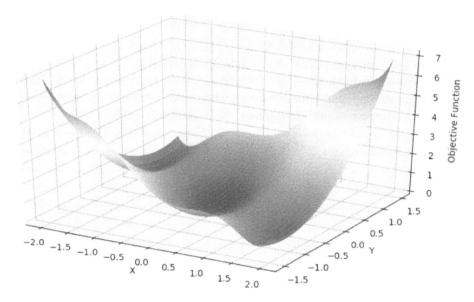

Figure 14.6 – Another example of a non-convex objective function showing multiple local minima

For optimization problems with non-convex objective functions, it is not easy to find the global minimum. No matter which algorithm we use, chances are that we will get a local minimum as the solution. There are, however, algorithms that perform iterative procedures to find good local minimum values. Random restart hill climbing and simulated annealing are examples of such algorithms. These iterative algorithms can be run on multiple machines or processors – not only to find a good local optimum solution in a short amount of time but also to be able to search multiple locations in the search space for the objective function concurrently.

In the next section, we will discover some of the commonly used numerical optimization algorithms.

Common numerical optimization algorithms

Several numerical optimization algorithms are implemented in open source and commercially sold optimization software tools. A lot of these algorithms are based on heuristic search, which is a technique based on solving problems quickly compared to classic methods. Heuristics-based algorithms attempt to find an approximate solution since the exact solution is very hard to find. The solutions provided by heuristics-based methods are considered good enough to solve the problem; however, it is generally not the best solution. In this section, we will briefly discuss a few of these algorithms. For detailed discussions on these algorithms and their mathematical formulation, you can refer to the articles and texts cited in the *Further reading* section of this chapter.

Random restart hill climbing

In **hill climbing**, we start from a point, $x \in X$, and search in the neighborhood of x. If the value of the objective function, f, increases in any direction in the neighborhood of x, then we move in the direction of the increment. We stop when the value of the objective function does not increase in any direction. This is the local optimum value of the objective function relative to our starting point. This method is also called **steepest ascent hill climbing**. The algorithm is very simple:

$$Let \; n^* = \underset{n \in N(x)}{\operatorname{argmax}} f(n)$$

$$if \; f(n^*) > f(x): x = n$$

$$else: stop$$

Figure 14.7 shows an example of hill climbing to a local maximum value:

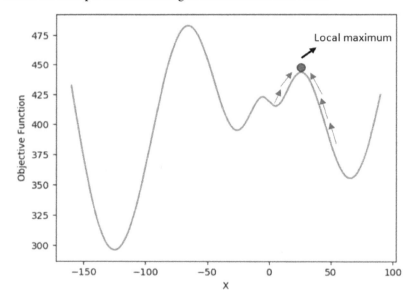

Figure 14.7 – Example of hill climbing to a local optimum in the objective function

For the case of minimization problems, the algorithm searches for a valley (or local minimum value) in the objective function. Note that the optimum value found in *Figure 14.7* is a local optimum value and it depends on where we started the search from.

Random restart hill climbing is an extension of the hill climbing method, in which, after finding an optimum value, the algorithm starts again at a different location in the variable space. This will often result in arriving at a different optimum value, as shown in *Figure 14.8*:

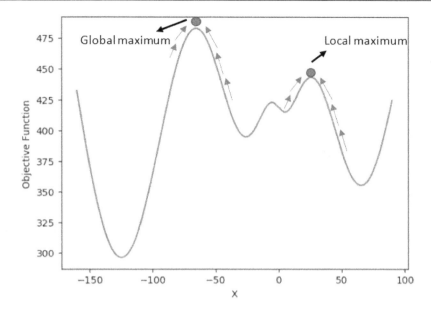

Figure 14.8 – Random restart hill climbing starting at two different variable values
and then using hill climbing to get to the optimum value in the vicinity

Even though the algorithm arrives at the global optimum value in *Figure 14.8*, this will not always be the case. However, if we run random restart hill climbing several times, chances are that our resulting local optimum value at the end will be better than just trying hill climbing only once. Since random restart hill climbing starts with different values of the variables, which are chosen randomly each time, each iteration can be run on separate processors and threads to speed up the algorithm.

Simulated annealing

In random restart hill climbing, we only move in one direction – the direction toward a local maximum (or minimum, depending on the problem type). This means that the algorithm only exploits the information and is not exploring outside of its immediate neighborhood. By not exploring at all, there is a good chance that the algorithm will get stuck in a local optimum value and will stay there. In **simulated annealing**, the algorithm also explores. It is not always trying to improve upon the current objective function value (to move in the direction of the local optimum), but it also sometimes moves in the direction where the objective function value gets worse (opposite to the direction of the local optimum). The simulated annealing algorithm is described as follows.

For a finite set of iterations, do the following:

1. Sample the new point, x_t, in the neighborhood, $N(x)$, of the current point, x

2. Jump to the new point with the probability given by an acceptance probability function, $P(x, x_t, T)$, where T is the temperature parameter that controls how often we jump, and f is the objective function value:

$$P = \begin{cases} 1 & \text{if } f(x_t) \geq f(x) \\ e^{\frac{f(x_t)-f(x)}{T}} & \text{otherwise} \end{cases}$$

3. Decrease temperature, T $(T > 0)$

In the preceding expression, if the objective function value for the new point is greater than the current value, then we make the jump to the new point. If the objective function value for the new point is less than the current value, we make the jump to the new point with the following probability:

$$e^{\frac{f(x_t)-f(x)}{T}}$$

Now, let's look at the effects of temperature on simulated annealing.

Effects of temperature, T

The following are the effects of temperature, T, in simulated annealing:

*   If T is large, the exponential will be close to 1, and we would make the jump with high probability, regardless of the objective function value at the new point, x_t. This is very similar to random walk when $T \to \infty$.

*   If T is small, the exponential will be very small and we would rarely make the jump to the new point. This is very similar to hill climbing when $T \to 0$.

During the algorithm run, T is generally decreased slowly. When T is large, we jump around in the objective function space quite often and there is a good chance that we will end up somewhere close to the global optimum value or a good local optimum value for the objective function. By the time we have reduced T to a small value, we are probably very close to the global optimum and hence looking only in its vicinity. The probability of ending at a point, x, is given as follows:

$$Pr(ending\ at\ x) = \frac{e^{\frac{f(x)}{T}}}{z_T}$$

Here, z_T scales the probabilities between 0 and 1. As we can see from this expression, the larger the value of the objective function (in the case of the global maximum), the larger the probability that we will end at that point, x. The same holds for the case of minimization tasks.

Figure 14.9 demonstrates the concept of simulated annealing, along with the probabilities of moving to the new point, depending on whether the newly selected points improve the objective function or not:

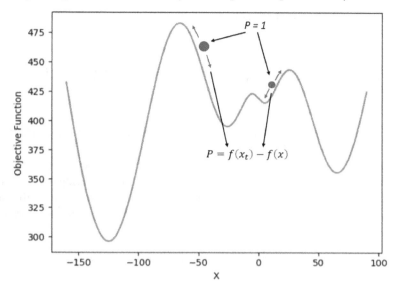

Figure 14.9 – Probability of moving to a new point in simulated annealing
when a neighboring point is selected to be the next point

Let's discuss Tabu search next.

Tabu search

Tabu search is another heuristic-based numerical optimization method conceptually similar to simulated annealing. Just like simulated annealing, we are allowed to move to a solution where our objective function value worsens. In Tabu search, local search is carried out and it is prohibited to come back to previously visited solutions. A Tabu list is maintained that consists of rules and solutions that are not allowed to be explored during the local search, giving this method the name Tabu search.

Evolutionary methods

Evolutionary algorithms are population-based algorithms that use candidate solutions along with some fitness function to evolve/improve the solution using mutation and recombination. Evolutionary methods are used quite often in numerical optimization problems and can find a good local optimum generally within a few iterations. Genetic algorithms are a very well-known and used class of evolutionary algorithms. Genetic algorithms have several applications in the domain of numerical optimization, as well as machine learning. They can be used with binary as well as non-binary representations.

Genetic algorithms generally use two solutions and apply a crossover operator to these solutions to arrive at a better solution.

In the next section, we will discuss the various applications and use cases of large-scale numerical optimization problems.

Example use cases of large-scale numerical optimization problems

In the previous section, we discussed a few of the commonly used numerical optimization methods. There are several others that we did not touch upon, and we recommend you check the *References* section for some great texts on several numerical optimization methods. Several very common large-scale optimization problems are implemented and solved in verticals, such as logistics, manufacturing, telecommunications, health care and life sciences, financial services, and so on. In this section, we are going to discuss a few of the very common practical large-scale numerical optimization use cases and applications. We will discuss the following use cases:

- The traveling salesperson problem of determining the best route for a salesperson going from one city to the next

- A dispatch optimization problem for technicians traveling via vehicles and carrying out various jobs in a geographic location

- Assembly line optimization to allocate the optimal type and number of parts to be manufactured on an assembly line

We will begin by discussing one of the oldest and most commonly studied numerical optimization problems, known as the traveling salesperson problem.

Traveling salesperson optimization problem

The traveling salesperson problem is one of the most studied combinational optimization problems, first formulated in 1930. It belongs to the class of NP-hard problems; the decision version of this problem belongs to the class of NP-complete problems. In the traveling salesperson problem, we are given a set of cities (or locations), and we start from a city, travel to each city exactly once, and return to the origin city to find the shortest route to accomplish this task. For example, as shown in *Figure 14.10*, in the US map, we want to start from city A, travel through all the cities marked, and then return to city A while following the shortest possible route:

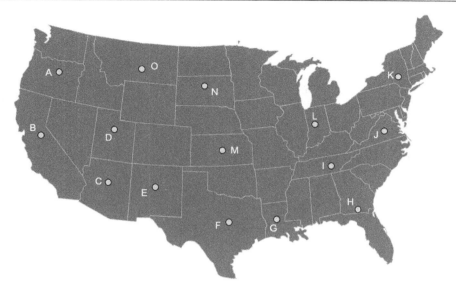

Figure 14.10 – A map of the US showing arbitrary cities A through O

Even though this problem seems simple to solve, it is an NP-hard problem. There are several combinations that the traveling salesperson can take to visit each city exactly once, but there is generally only one solution that accomplishes this with the shortest possible route. The traveling salesperson problem can be formulated in a few different ways. It can be formulated as an undirected weighted graph with cities being the vertices of the graph, the route connecting the cities being the edges, and weighted by the distance between the cities. It can then be solved as a minimization problem that starts and finishes at a given vertex, with each vertex being visited exactly once.

Another way to model the traveling salesperson problem is as an integer linear program. Several formulations can be used, such as the Miller-Tucker-Zemlin formulation and the Dantzig-Fulkerson-Johnson formulation. When the number of cities is small and only a small set of paths exists between the cities, the exact solution can be found in a small amount of time. However, as the number of cities and routes between the cities become large, finding the exact solution in a reasonable amount of time becomes almost impossible. In such situations, numerical methods attempt to find approximate or suboptimal solutions for the problem. *Figure 14.11* shows one such example of a route found between the cities shown in *Figure 14.10*. There may be possible shortest routes that exist for this problem, but the route shown in *Figure 14.11* is reasonably good and is quite possibly either the shortest possible route or very close to the shortest possible route:

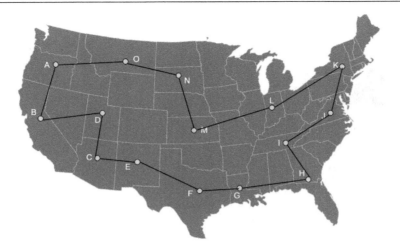

Figure 14.11 – Example of the traveling salesperson problem showing a very good route (possibly the shortest possible route) between the cities marked from A through O on the map

The traveling salesperson problem has several applications in various fields, such as planning, logistics, manufacturing, DNA sequencing, and so on. There are also several open source and commercially sold software tools available for finding a solution for the traveling salesperson problem, as well as extending it to additional similar problems that exist in the industry. One of the most common practical extensions of the traveling salesperson problem is the vehicle routing problem. In the following section, we will discuss a more complex extension of the vehicle routing problem, called the worker dispatch optimization problem.

Worker dispatch optimization

The vehicle routing problem attempts to find the optimal set of routes for a fleet of vehicles to make deliveries to customers. This is a very common problem in logistics where organizations such as the **United States Postal Service (USPS)**, **United Parcel Service (UPS)**, FedEx Corporation, and others, have to deliver several packages to a set of customers in various geographic locations daily. This problem can be formulated using various business objectives, such as delivering the packages promptly while maximizing the number of deliveries assigned to a driver/vehicle and minimizing the total fuel cost. These organizations generally formulate this problem as an optimization problem and then solve it daily (and often multiple times a day) to make the delivery to customers while also maximizing their profitability within certain constraints.

Similar to the traveling person problem, finding a solution for the vehicle routing problem is also NP-hard. However, several numerical optimization software tools find a very good local optimum solution in a reasonable amount of time. Furthermore, using high-performance and distributed computing, the software can be written to start searching for various solutions at the same time on multiple processors and machines and then aggregate and find the best one out of the various local optima found.

Often, the vehicle routing problem is also extended with a few modifications to solve even more complex problems. One such example is the technician or worker dispatch optimization problem. In a worker dispatch optimization problem, the goal is to send technicians or workers to a customer location and carry out some task that requires time to complete. This is a very common problem for service-providing organizations such as electricity, gas, internet, telecommunication, and so on. These organizations have several worker/technician hubs or garages based on the home location of the worker. The jobs that arrive each day need to be assigned to these workers based on their schedules, as well as their skills and skill competency levels, since not all jobs are always the same. All workers have the same skill level for different types of jobs. Furthermore, in all such jobs, there is a committed time window to the customer that needs to be met for customer satisfaction. In addition, different jobs take a different amount of time to complete, and there is also the time required to travel a distance to the customer location, which may vary based on the time of day.

Figure 14.12 shows an example where there is a worker hub in the center and several customer locations that need to be serviced by workers/technicians on a given day:

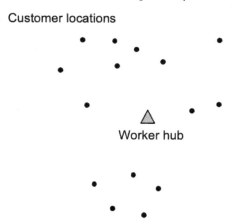

Figure 14.12 – An example of the worker/technician optimization problem

We will now outline these constraints formally and also discuss a few objective functions that can be used to solve this problem.

Possible objective metrics for the worker optimization problem

The worker optimization problem can have several objective functions/metrics based on the requirements of the business. We will list a few of these here:

- **Minimize total fuel cost**: With this objective function, the goal is to find a non-trivial solution that minimizes the total fuel cost per job of all the vehicles on any given day. Fuel costs will depend on the number of workers and vehicles, the number of jobs, the location of jobs, the routes that were taken to get to the jobs, the order of the jobs, and the worker's schedules. Often, minimizing fuel costs is an indirect consequence of finishing the maximum number of jobs on a given day, because if the algorithm can pack a large number of jobs for a given worker, then the fuel cost per job for that particular vehicle will be low and hence the total fuel cost per job will be low as well for all the workers.

- **Maximize the number of jobs done on a given day**: To maximize profit and also to keep the customers happy, service organizations need to maximize the number of jobs completed on a given day with as few jobs as possible being carried over to the next day. This also depends on the number of workers available on a given day, worker skills, and the route taken to carry out the jobs.

- **Maximize worker efficiency**: This objective is dependent on maximizing the number of jobs done on a given day. The goal of this objective function is to maximize the number of jobs carried out by each worker on a given day, which also depends on the technician's schedule, skills, level of competency, starting location, and the distance needed to travel by the technician.

- **Composite objective function**: In a composite objective function, the goal is to explicitly optimize a combination of objective functions, such as maximizing worker efficiency and minimizing distance or fuel cost. Sometimes, the various terms in a composite objective function may also be opposite to each other. For example, increasing one may result in decreasing the other, and so on. In such cases, we may have penalty terms associated with the different objective functions comprising the composite function and optimize the resulting combination of penalized terms.

Now, let's look at some of the constraints that can be important for formulating the worker dispatch optimization problem.

Important constraints for the worker dispatch optimization problem

The following are some of the important constraints for the worker dispatch optimization problem. While constraints cover the most common ones, there can be additional constraints, depending on the individual use case:

- **The number of jobs**: The total number of jobs for a given worker hub on any given day is important and affects the worker efficiency number, as well as the distance traveled, and hence the fuel cost. How many jobs can be feasibly completed on any given day is also heavily dependent on the number of jobs on a given day.

- **The number of workers available on a given day**: How many workers are available on a given day to carry out the jobs is another important constraint and it has a significant impact on worker efficiency, as well as distance traveled and fuel cost.

- **Worker schedule**: In addition to the number of workers available, the schedule of each worker is also an important constraint. Some workers may start their shift at 8 A.M. and some at 10 A.M., and so on. Similarly, the number of hours each worker may work each day might be different. Some may work for 8 hours, and others for 6 hours. In addition, generally, workers also have break times, such as lunch and other periodic breaks. These breaks may also be at different times, adding further schedule-related constraints to the optimization problem.

- **Job types**: There are generally several different job types. For example, for a telecommunication organization, there might be new service installations or old service repair jobs. In addition, several service-providing organizations offer multiple products and services. For example, a telecommunication organization generally offers internet, cable/TV services, and home phone services. These different job types add another dimension to the optimization problem, further complicating it.

- **Worker skills and skills-related competency levels**: Just like the different job types mentioned previously, different workers and technicians have different skill types as well as expertise levels in the skill. Using the same telecommunication use case example again, some workers might be dedicated to installing new services and others to repairing old services. In addition, some technicians can be experts in internet service installation and others in telephone service installation. This results in different workers taking different amounts of time to install the same service or debug and repair the same problem. This also adds an interesting dilemma when formulating the problem from a business point of view.

  The business may want to maximize the total number of jobs completed on a given day, which is generally accomplished if the algorithm matches the skill levels of workers with the jobs appropriately. On the other hand, if the business follows this approach, then the workers may not be able to learn new skills or get practical experience related to services and problems that they are not already an expert at. This parameter regarding skill level should be modeled appropriately in the formulation of the problem to achieve the best results.

- **Job locations**: Where each job is located in a geographical location is also important for deciding the route assigned by the optimization solution to each worker.

- **Customer time windows**: Service-providing companies also commit to a specific time window in which the worker should arrive at the customer/job location. These time windows may also vary based on the type of job, the number of workers available, as well as geographical locations. These time windows also have a significant effect on the final objective function value. For example, there might be a new service install job at a customer location with a committed time window of 8–10 A.M. on a particular day. At the same time, there might be another customer very close by with a repair request. Now, even though the jobs are physically very close to each other, because of the promised time window, the organization will probably need to dispatch multiple workers to adhere to the times committed to the customer. Due to this, several modernized organizations are also formulating the worker dispatch problem jointly with the scheduling problem; when there is a request for a repair or an installation, the scheduler should take into account all these constraints while committing to a time window with the customer.

- **Job durations**: Different jobs take different amounts of time to complete. There is generally an average time for a particular job for all the workers in a worker hub, and also, individual times are taken by each worker to complete that job. All these are also modeled as constraints in the optimization problem to get the best results.

- **Maximum travel time and distance**: Generally, there is also a maximum limit on how much total distance or time a worker may travel on a given day, as well as the farthest a worker may travel from the garage hub.

In addition to these constraints, there may be additional constraints too (for example, the weather: rain, snow, storm, and so on), depending on the particular organization working on the use case. As we can imagine, all these constraints make the worker dispatch optimization problem very complicated. Generally, for any organization using this approach to assign jobs to its workers, this problem is solved on multiple higher-performance computation machines every morning in a distributed fashion. For example, for the same geographic location, the random restart approach for hill climbing and other similar algorithms can be used, with each restart iteration being executed on a different processor and/or machine. There are several open source and commercially available optimization software that formulate and solve this problem very efficiently in a reasonable amount of time.

Figure 14.13 shows an example of three worker hubs with three workers each following an optimized route to the job locations and back to the worker hub:

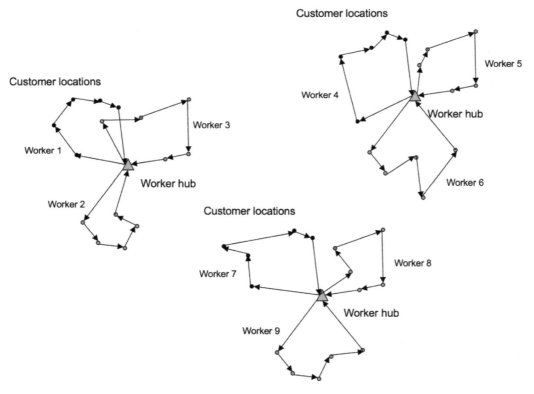

Figure 14.13 – Example showing three worker hubs and nine workers in total, with three workers each starting from a worker hub and following an optimized route to the job locations

By using numerical optimization algorithms to solve this optimization problem, service companies can improve worker efficiency, cut down on fuel costs, and improve customer satisfaction significantly. Next, we will discuss another example of numerical optimization for allocating items to an assembly line to maximize the number of produced items on a given day.

Assembly line optimization

In manufacturing industries such as electronics manufacturing, there are usually several assembly lines or belts on which various items are being built and assembled to build a final product, such as a desktop/laptop computer, cell phone, tablet, and so on. On these assembly lines, human workers are manually working on assembling the items. Different assembly lines can assemble different products, with some overlap. Furthermore, the workers assembling the products also vary in some skills and skill levels, just like the worker dispatch optimization problem. Let's discuss the various optimization metrics that can be used for this problem based on the business use case.

Objective metrics for assembly line optimization

The following objective metrics are some of the common ones used for the assembly line optimization problem:

- **Maximize the number of items produced on a given day**: Using this metric, the goal is to maximize the total number of items assembled on any given day. Generally, it also depends on the number of orders, as well as the forecast for the number of items needed in near future.

- **Minimize the number of items stored in storage**: With this metric, the goal is to minimize the number of excessive items manufactured and stored in the storage for future sales. This metric also depends on the forecast of items to be manufactured, as well as the storage capacity.

In addition to these metrics, other metrics can be used based on the business use case and goals. Furthermore, like the worker dispatch optimization problem, objective metrics comprised of multiple metrics can also be used. Let's look at some of the constraints for this problem.

Constraints for the assembly line optimization problem

The following constraints are important for the assembly line optimization problem:

- The number of different items needed to be assembled each day. This constraint depends on the sales forecast, as well as the number of pre-ordered items.

- The number of workers available to work on a given day, as well as the schedule of each worker.

- The skills of various individual workers, as well as the skill competency level of each worker.

- Skills needed to assemble various items.

- The capacity of each belt to assemble different items.

- Storage capacity of the factory and/or warehouse.

- The maximum number of items that can be stored (surplus) for a specific amount of time, such as a day, a week, or a month.

While these are some of the common constraints being considered while formulating the assembly line optimization problem, there can be additional constraints as well, depending on the specific business use case and various other conditions and requirements. By formulating this problem as an optimization problem and then solving it on a daily, weekly, monthly, or quarterly schedule, manufacturing companies generally improve their yield, profit, and efficiency while reducing waste and the number of surplus items.

In this section, we discussed a few applications and use cases of numerical optimization being used in the industry. In the next section, we are going to discuss the high-performance compute options available on AWS for solving these numerical optimization problems.

Numerical optimization using high-performance compute on AWS

As discussed in the previous sections, most of the numerical optimization problems are NP-hard and highly compute-intensive for finding a reasonable solution. The software tool employing these algorithms has to carry out a large-scale search over a very complicated multidimensional objective function in search of the global optimum. Because of the complexity, the number of dimensions, non-convexity, and sometimes discontinuities present in these objective functions, it is almost impossible to find the global optimum in a finite amount of time, even with today's compute resources.

However, for most of these problems, several commercially available and open source software tools find a very good solution (local optimum) in a reasonable amount of time. These tools can be run on the infrastructure and compute resources provided by AWS. Let's discuss some of the common commercial and open source tools that can be installed and run on various AWS resources to solve numerical optimization problems in almost every industry domain.

Commercial optimization solvers

The following commercial solvers are some of the most popular and common ones used on AWS compute infrastructure:

- IBM ILOG CPLEX Optimization Studio (commonly known as **CPLEX**)

- Gurobi Optimization

- FICO Xpress Optimization

- **A Mathematical Programming Language** (**AMPL**)

Open source optimization solvers

In addition to the commercially sold optimization solver tools, the following open source optimization solvers can be easily run on AWS compute infrastructure as well:

- **GNU Linear Programming Kit (GLPK)**
- **Computational Infrastructure for Operations Research (COIN-OR)**
- Pyomo
- **Convex Over and Under ENvelopes for Nonlinear Estimation (Couenne)**
- PuLP
- Google OR-Tools
- SCIP Optimization Suite

These commercially available and open source software tools can be run on AWS infrastructure using a variety of different architecture patterns, as outlined in the following section.

Numerical optimization patterns on AWS

Various architecture patterns can be employed to run the previously mentioned optimization software tools on AWS resources.

Figure 14.14 shows various tools and resources from the AWS stack that can be used to help with solving numerical optimization problems:

Figure 14.14 – Various AWS resources and tools that can be used to solve numerical optimization problems

Let's discuss a few architecture patterns employing these AWS resources and tools and see how they can help with solving numerical optimization problems.

EC2 instances

We can install and run these optimization tools on Amazon EC2 instances in a container. The optimization software suite, along with all the required libraries, can be built into a container that can then utilize EC2 instances, which can also be used in a distributed manner to run several parallel searches at the same time (for example, random restart hill climbing). By running multiple iterations of these algorithms in parallel, there is a better chance of arriving at the global optimum value or a very good local optimum. *Figure 14.15* shows the architecture for running these optimization tools on EC2 instances using containers:

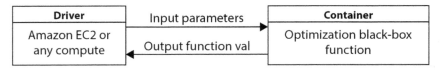

Figure 14.15 – Example architecture showing numerical optimization software
running on a container on an Amazon EC2 compute instance

Using a serverless architecture

In addition to using EC2 instances, we can also run the optimization software in a serverless manner on AWS. One example of using a serverless architecture is shown in *Figure 14.16*, where AWS Lambda is used to launch multiple optimization tasks on AWS Fargate. These tasks can be run in parallel and then aggregated to get the best solution for an optimization problem. These tasks can also be different optimization packages and libraries attempting to solve the same problem, with the best result being used at the end. The data consisting of constraints and variables can be read from Amazon S3, as also shown in *Figure 14.16*. Amazon CloudWatch is also used in this pattern to output the necessary steps and status messages:

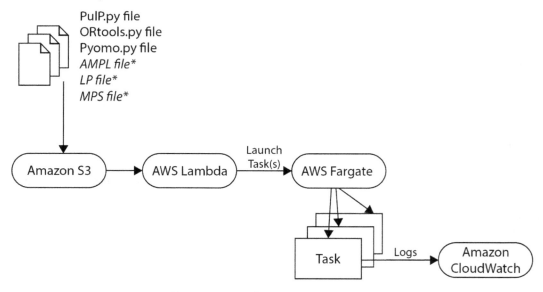

Figure 14.16 – An example of a serverless architecture using AWS Lambda
and AWS Fargate to run various optimization tasks in parallel

The advantage of using this approach over that of EC2 instances is cost and scalability. We can start as many optimization tasks as needed without having to worry about managing EC2 instances. In addition, since we are using a serverless architecture, we only need to pay for the compute for the duration that our optimization tasks are running. As an example, for our worker dispatch optimization problem, the job and worker-related data can arrive in an Amazon S3 bucket every morning. Then, using AWS Lambda, various optimization tasks can be launched on AWS Fargate, with each task for a particular worker hub attempting to find the optimal route and schedule for every worker in the worker hub.

Using Amazon SageMaker processing

In addition to using dedicated EC2 instances and a serverless architecture, we can also carry out numerical optimization on Amazon SageMaker using SageMaker processing jobs. *Figure 14.17* shows an example of this architecture pattern, where data consisting of constraints and various variables is residing in an Amazon S3 bucket. AWS Lambda and an AWS Step function are used to launch a SageMaker processing job that reads the data from the S3 bucket and runs the optimization task in a container with all the required packages and software needed to run the optimization job. This optimization job is run on an ephemeral EC2 instance; once the job is completed, the instance is released and no more cost for the instance is incurred. The results are written in S3 and also in Amazon DynamoDB after some post-processing by an AWS Lambda function. A few other AWS resources are shown in *Figure 14.17*, such as for authentication and caching, which may or may not be necessary, depending on the specific use case:

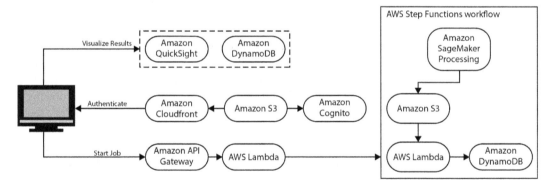

Figure 14.17 – An example of running numerical optimization using Amazon SageMaker processing

In this section, we looked at a few ways numerical optimization problems can be solved using AWS high-performance compute resources and tools. While these are good examples of architectural patterns that can be employed for a variety of use cases, these can also be modified and extended, depending on the use case and business requirements.

In the next section, we are going to look at how numerical optimization is important for solving machine learning problems as well.

Machine learning and numerical optimization

So far, we have discussed numerical optimization and its use cases from an optimization problems perspective. Whereas numerical optimization has several standalone industry use cases and applications, it is also very commonly used in several machine learning algorithms and use cases. Whether it's supervised learning, unsupervised learning, or reinforcement, we are always solving some form of optimization problem using iterative processes at the very core of a machine learning algorithm.

In supervised learning, for example, let's look at the case of linear regression. In linear regression, we are minimizing a cost function consisting generally of the mean squared error between the actual value of a target variable and the value predicted via the model.

Our algorithm arrives at the minimum value of the cost function (convex function with a global minimum if it is mean squared error, non-convex with local minima in most other cases) using an iterative algorithm, such as gradient descent. Gradient descent looks at the gradient of the cost function and then modifies the linear regression parameters in the direction of the gradient. This way, after a certain number of iterations, the algorithm arrives at the global or local minimum value of the objective function.

Similarly, in logistic regression, we use an objective function consisting of logarithm terms. This objective function is again convex and we use gradient descent again to arrive at the minimum value of the objective function. Hence, once again, we are solving a numerical optimization problem at the core of the problem, with the higher-level goal of building a machine learning model for a classification problem.

In neural networks, including deep neural networks, we have several parameters for which we need to find the optimal value so that some error is minimized at the output layer. In the field of deep learning, we often have very large machine learning models consisting of millions or even billions of weights or parameters, especially for natural language processing and computer vision problems. Each neuron or unit in the neural network has some activation function, which is a function of a few of these parameters. To build a model that fits the data well and does good predictions on tests or new data, we need to find the optimal value of neural network parameters/weights. This, again, is achieved by using gradient descent or some other optimization algorithm to solve this numerical optimization problem consisting of a very large number of parameters.

Similar to supervised learning, numerical optimization is also used in unsupervised learning problems. For example, in clustering methods such as K-means clustering, we are trying to minimize the distance between the cluster center and associated points in the cluster. Similarly, in expectation maximization (a soft clustering method), we are maximizing the likelihood of each data point being generated by a Gaussian distribution, whose mean we are attempting to find.

In reinforcement learning, numerical optimization is also quite often used. For example, in most reinforcement learning methods, the goal of the algorithm is to maximize some form of long-term reward using actions and rewards, while simulating the scenario repeatedly to learn the optimal policy that maximizes long-term reward. In deep reinforcement learning, we use the neural network weights to approximate the policy. These weights are, again, learned using numerical optimization algorithms such as gradient descent. In short, no matter which type of machine learning problem we are trying to solve, there is a strong connection with numerical optimization, and for most of these machine learning problems, we are solving some form of optimization problem to get the answer for our machine learning problem.

Let's summarize what we've learned in this chapter.

Summary

In this chapter, we discussed numerical optimization and its applications. We started with a discussion about numerical optimization and its necessary ingredients. Next, we discussed a few of the common numerical optimization methods. We also discussed a few large-scale applications and use cases of numerical optimization. These use cases are very well known in academia as well as in the industry and are implemented by several organizations in their businesses. In addition, we talked about how AWS high-performance compute options and resources can be used to solve numerical optimization methods, and also discussed a few architectural patterns to accomplish this.

Finally, we ended with a short discussion about how various categories of machine learning algorithms employ numerical optimization at their core to build good models. The topics covered in this chapter will help you understand and formulate numerical optimization use cases, how numerical optimization is important for machine learning, and how high-performance computing can help with solving numerical optimization use cases. In addition, you should have an idea of the tools and software available for solving numerical optimization problems.

Overall, in this book, we have discussed the fundamentals of high-performance computing, followed by the data management, transfer, compute, networking, and storage aspects of high-performance computing. We also talked about applied modeling and its examples, such as data analysis, preprocessing, visualization, distributed training of machine learning models, optimizing models and their deployment, along with scaling machine learning models. Furthermore, we looked at various applications of high-performance computing, such as computational fluid dynamics, genomics, autonomous vehicles, and numerical optimization. The material presented in this text will introduce you to all these concepts and enable you to explore further and solve interesting use cases in high-performance computing and associated fields.

Further reading

To learn more about the topics that were covered in this chapter, take a look at the following resources:

- *An Interactive Tutorial on Numerical Optimization*: `https://www.benfrederickson.com/numerical-optimization/`
- El-Ghazali Talbi. 2009. *Metaheuristics: From Design to Implementation*. Wiley Publishing.

- Refs on NP-hard and completeness:

  - The traveling salesperson problem: `https://en.wikipedia.org/wiki/Travelling_salesman_problem`

  - The vehicle routing problem: `https://www.sciencedirect.com/topics/economics-econometrics-and-finance/vehicle-routing-problem`

  - IBM ILOG CPLEX Optimization Studio: `https://www.ibm.com/analytics/cplex-optimizer`

  - Gurobi Optimization: `www.gurobi.com`

  - FICO Xpress Optimization: `https://www.fico.com/en/products/fico-xpress-optimization`

  - AMPL: `https://ampl.com/`

  - GNU Linear Programming Kit: `https://www.gnu.org/software/glpk/`

  - Computational Infrastructure for Operations Research: `https://www.coin-or.org/`

  - Pyomo: `http://www.pyomo.org/`

  - Convex Over and Under ENvelopes for Nonlinear Estimation: `https://github.com/coin-or/Couenne`

  - PuLP: `https://pypi.org/project/PuLP/`

  - Google OR-Tools: `https://developers.google.com/optimization`

  - SCIP Optimization Suite: `https://www.scipopt.org/`

Index

Packt.com

Subscribe to our online digital library for full access to over 7,000 books and videos, as well as industry leading tools to help you plan your personal development and advance your career. For more information, please visit our website.

Why subscribe?

- Spend less time learning and more time coding with practical eBooks and Videos from over 4,000 industry professionals
- Improve your learning with Skill Plans built especially for you
- Get a free eBook or video every month
- Fully searchable for easy access to vital information
- Copy and paste, print, and bookmark content

Did you know that Packt offers eBook versions of every book published, with PDF and ePub files available? You can upgrade to the eBook version at packt.com and as a print book customer, you are entitled to a discount on the eBook copy. Get in touch with us at customercare@packtpub.com for more details.

At www.packt.com, you can also read a collection of free technical articles, sign up for a range of free newsletters, and receive exclusive discounts and offers on Packt books and eBooks.

Other Books You May Enjoy

If you enjoyed this book, you may be interested in these other books by Packt:

Accelerate Deep Learning Workloads with Amazon SageMaker

Vadim Dabravolski

ISBN: 978-1-80181-644-1

- Cover key capabilities of Amazon SageMaker relevant to deep learning workloads
- Organize SageMaker development environment
- Prepare and manage datasets for deep learning training
- Design, debug, and implement the efficient training of deep learning models
- Deploy, monitor, and optimize the serving of DL models

Machine Learning Engineering on AWS

Joshua Arvin Lat

ISBN: 978-1-80324-759-5

- Find out how to train and deploy TensorFlow and PyTorch models on AWS
- Use containers and serverless services for ML engineering requirements
- Discover how to set up a serverless data warehouse and data lake on AWS
- Build automated end-to-end MLOps pipelines using a variety of services
- Use AWS Glue DataBrew and SageMaker Data Wrangler for data engineering
- Explore different solutions for deploying deep learning models on AWS
- Apply cost optimization techniques to ML environments and systems
- Preserve data privacy and model privacy using a variety of techniques

Packt is searching for authors like you

If you're interested in becoming an author for Packt, please visit `authors.packtpub.com` and apply today. We have worked with thousands of developers and tech professionals, just like you, to help them share their insight with the global tech community. You can make a general application, apply for a specific hot topic that we are recruiting an author for, or submit your own idea.

Share Your Thoughts

Now you've finished *Applied Machine Learning and High-Performance Computing on AWS*, we'd love to hear your thoughts! Scan the QR code below to go straight to the Amazon review page for this book and share your feedback or leave a review on the site that you purchased it from.

`https://packt.link/r/1-803-23701-5`

Your review is important to us and the tech community and will help us make sure we're delivering excellent quality content.

Download a free PDF copy of this book

Thanks for purchasing this book!

Do you like to read on the go but are unable to carry your print books everywhere? Is your eBook purchase not compatible with the device of your choice?

Don't worry, now with every Packt book you get a DRM-free PDF version of that book at no cost.

Read anywhere, any place, on any device. Search, copy, and paste code from your favorite technical books directly into your application.

The perks don't stop there, you can get exclusive access to discounts, newsletters, and great free content in your inbox daily

Follow these simple steps to get the benefits:

1. Scan the QR code or visit the link below

https://packt.link/free-ebook/978-1-80323-701-5

2. Submit your proof of purchase
3. That's it! We'll send your free PDF and other benefits to your email directly

www.ingramcontent.com/pod-product-compliance
Lightning Source LLC
Chambersburg PA
CBHW062047050326
40690CB00016B/3005